Collins

AQA GCSE
Maths

Foundation Skills Book

Reason, interpret and communicate
mathematically and solve problems

Sandra Wharton

William Collins' dream of knowledge for all began with the publication of his first book in 1819. A self-educated mill worker, he not only enriched millions of lives, but also founded a flourishing publishing house. Today, staying true to this spirit, Collins books are packed with inspiration, innovation and practical expertise. They place you at the centre of a world of possibility and give you exactly what you need to explore it.

Collins. Freedom to teach

Published by Collins
An imprint of HarperCollins*Publishers*
The News Building
1 London Bridge Street
London SE1 9GF

> Browse the complete Collins catalogue at
> www.collins.co.uk

© HarperCollins*Publishers* Limited 2015

10 9 8 7 6 5 4 3 2 1

ISBN 978-0-00-811386-5

A catalogue record for this book is available from the British Library

The author Sandra Wharton asserts her moral rights to be identified as the author of this work.

Commissioned by Lucy Rowland and Katie Sergeant
Project managed by Elektra Media Ltd and Hart McLeod Ltd
Contributions from Jo-Anne Lees and Brian Speed
Copyedited by Jim Newall
Proofread by Joanna Shock
Answers checked by Jim Newall
Edited by Caroline Green and Jennifer Yong
Typeset by Jouve India Private Limited
Illustrations by Ann Paganuzzi
Designed by Ken Vail Graphic Design
Cover design by We are Laura
Production by Rachel Weaver

Printed in Italy by Grafica Veneta S.p.A.

Acknowledgements
The publishers gratefully acknowledge the permissions granted to reproduce copyright material in this book. Every effort has been made to contact the holders of copyright material, but if any have been inadvertently overlooked, the publisher will be pleased to make the necessary arrangements at the first opportunity.

The publishers would like to thank the following for permission to reproduce photographs in these pages:

Cover (bottom) Georgios Kollidas/Shutterstock, cover (top) Godruma/Shutterstock.

Contents

How to use this book

Welcome to the *AQA GCSE Maths Foundation Skills Book*.

Focused on the new assessment objectives AO2 and AO3, this book is full of expertly written questions to build your skills in mathematical reasoning and problem solving.

This book is ideal to be used alongside the Practice Book or Student Book. It is structured by strand to encourage you to tackle questions without already knowing the mathematical context in which they sit – this will help prepare you for your exams.

Hints and tips

Some questions have a hint at the end of the chapter to get you started, but you should try to answer the question first, before looking at the hint. Take your time with the longer and multi-step questions – they are designed to make you think!

Worked exemplars

These give you suggested ways of working through these sorts of questions, step by step.

Glossary

Explanations for the important words you need to know can be found in the glossary at the back of the book.

Answers

The answers are available online at www.collins.co.uk/gcsemaths4eanswers.

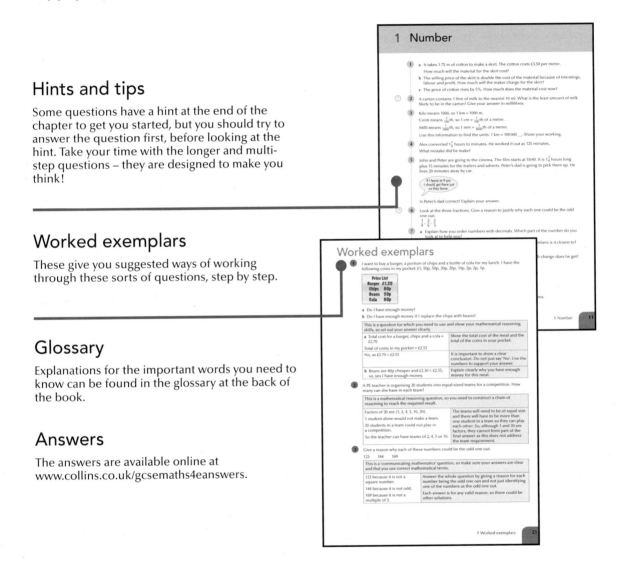

Teacher guide

- Build your students' confidence by tackling longer questions in class first, to generate discussions
- Questions are differentiated using a colour coded system from more accessible ⬤, through intermediate ⬤, to more challenging ⬤.
- Access the full mark scheme online

In the 2015 GCSE Maths specification there is an increased emphasis on problem solving, as demonstrated by the weighting of marks for each Assessment Objective (see below weightings for Foundation):

AO1: Use and apply standard techniques	50%
AO2: Reason, interpret and communicate mathematically	25%
AO3: Solve problems within mathematics in other contexts	25%

The questions in this Skills Book draw on content from each of the strands to help students become more confident in making connections between the different strands of mathematics that have previously often been seen in isolation.

Students should also be encouraged to identify types of questions so they can access the mathematics they need quickly and confidently.

There is an increased emphasis in the revised curriculum on the use of language and correct terminology with the expectation at all levels that the use of mathematical language should be more precise – the questions in this book have addressed this, and should help students to develop a strong ability to communicate mathematically.

The approaches used in this book build on research into how students learn effectively in mathematics by requiring them to:

- build on the knowledge they bring to each section
- engage with and challenge common misconceptions
- emphasise methods rather than answers
- create connections between mathematical topics.

Students are encouraged to develop their ability to think mathematically by:

- evaluating the validity of statements and generalisations through activities such as classifying statements as sometimes, always or never true
- being asked to interpret multiple representations
- classifying mathematical objects by using, for example, matching activities
- creating and solving new problems, and analysing reasoning and solutions.

To support students, some questions have hints to get them started. These are indicated by ⓐ. Build resilience by encouraging students to think carefully about the question before referring to these.

The Worked exemplars break down question types and show possible ways of tackling them in a step-by-step manner.

This book includes some longer and multi-step questions which start in a more straightforward way and get more challenging. Encourage students to take their time and think carefully about the information given to them, the answers required and the mathematics they will need to use.

Matching grid

Where questions relate to sections in the GCSE Maths Foundation Student Book, they are included in the grid below to make it easier to use the books alongside each other.

Foundation student book chapter and section	Question	Curriculum ref.
1 Number: Basic number		
1.1 Place value and ordering numbers	p11 Q7 , p15 Q33	N1
1.2 Order of operations and BIDMAS	p11 Q8 , p13 Q23, p31 Q38	N2, N3
1.3 The four rules	p11 Q1 , Q8 , p12 Q9 , Q14 , Q15 , p13 Q21 , Q24 , p14 Q28 , p15 Q33 , Q34 , p16 Q40	N2, N3
2 Geometry and measures: Measures and scale drawings		
2.1 Systems of measurement	p11 Q2 – Q5 , p12 Q16 , p17 Q50 , p19 Q57 , p41 Q1 , Q4 , p46 Q32	N13, R1, G14
2.2 Conversion factors	p15 Q37	N13
2.3 Scale drawings	p15 Q37 , p63 Q73	R2
2.4 Nets	p54 Q18 , p62 Q62	G13
2.5 Using an isometric grid	p62 Q62	G13
3 Statistics: Charts, tables and averages		
3.1 Frequency tables	p76 Q3 , Q4 , p80 Q23 , p83 Q35	S2
3.2 Statistical diagrams	p76 Q1 , p77 Q6 , Q8 , p78 Q10 , Q12 , Q14 , p79 Q15 , p81 Q24 , p83 Q30 , p84 Q38 , Q39	S2
3.3 Line graphs	p77 Q6 , p78 Q14 , p79 Q15 , p81 Q25	S2
3.4 Statistical averages	p76 Q1 , p77 Q7 , p78 Q13 , p80 Q16 , Q17 , Q20 , Q21 , p82 Q28 , Q29 , p83 Q35 , p84 Q40 , p85 Q41 , Q44	S4
4 Geometry and measures: Angles		
4.2 Triangles	p52 Q6 , p53 Q13 , p54 Q14 , Q17 , p57 Q26 , p61 Q53 , Q55	G1, G3, G4
4.3 Angles in a polygon	p57 Q28 , p61 Q49 , Q52 , Q54	G1, G3, G4
4.4 Regular polygons	p57 Q28 , p61 Q50 , Q54	G1, G3, G4
4.5 Angles in parallel lines	p57 Q27	G1, G3
4.6 Special quadrilaterals	p52 Q6 , p53 Q7 , Q8 , Q12 , p57 Q27 , p61 Q51 , Q56	G1, G3, G4
5 Number: Number properties		
5.1 Multiples of whole numbers	p12 Q12 , p13 Q26 , p14 Q28 , p16 Q45	N4
5.2 Factors of whole numbers	p12 Q11 , Q12 , p13 Q22 , Q25	N4
5.3 Prime numbers	p12 Q11 , Q12 , p14 Q27 , Q28 , p70 Q15	N4
5.4 Prime factors, LCM and HCF	p14 Q27	N4
5.5 Square numbers	p12 Q13 , p14 Q29	N6
5.6 Square roots	p17 Q48 , Q49	N6, N7
5.7 Basic calculations on a calculator	p12 Q10 , p30 Q34	N2, N3, N6, N7, N8, N9

Foundation student book chapter and section	Question	Curriculum ref.
6 Number: Approximations		
6.1 Rounding whole numbers	p11 Q2 , p15 Q36, p19 Q59	N15
6.2 Rounding decimals	p15 Q36, p16 Q42, Q43, p19 Q57, Q58, Q60, Q61, p63 Q71	N15
6.3 Approximating calculations	p13 Q22, p15 Q36, p18 Q54–Q56	N14, N15
7 Number: Decimals and fractions		
7.1 Calculating with decimals	p12 Q17, Q18, p13 Q21, Q22, p15 Q33, Q35, Q38, Q39, p16 Q41, p26 Q8 , p59 Q40, p60 Q44, Q45	N2, N14
7.2 Fractions and reciprocals	p11 Q6 , p12 Q19, Q20, p17 Q50, p33 Q52	N2, N10
7.3 Fractions of quantities	p11 Q4 – Q6 , p41 Q9 , p55 Q20	N2, N8, R3
7.4 Adding and subtracting fractions	p18 Q51, Q52, p41 Q9 , p42 Q12	N2, N8
7.5 Multiplying and dividing fractions	p13 Q24, p15 Q33, p42 Q12	N2, N8
7.6 Fractions on a calculator	p18 Q51, Q52	N2, N8
8 Algebra: Linear graphs		
8.1 Graphs and equations	p25 Q5 , p27 Q13, p28 Q26, Q29	A8, A9
8.2 Drawing linear graphs by finding points	p25 Q5	A8, A9
8.3 Gradient of a line	p35 Q57	A10
8.4 $y = mx + c$	p26 Q7 , p27 Q17, p28 Q22–Q24	A8, A9
8.5 Finding the equation of a line from its graph	p33 Q54	A10
8.6 The equation of a parallel line	p35 Q57	A9, A10
8.7 Real-life uses of graphs	p28 Q27, p29 Q30, Q31, p33 Q54	A14
9 Algebra: Expressions and formulae		
9.1 Basic algebra	p25 Q1 – Q4 , p26 Q8 –Q10, p27 Q11, Q12, Q14, Q16, Q18, Q19, p31 Q38, Q39, Q42, p33 Q53	A1, A3, A6
9.2 Substitution	p26 Q7 , p27 Q15, p28 Q22, Q24, p29 Q31, p31 Q40, Q41, Q44, p33 Q53, p35 Q55	A6
9.3 Expanding brackets	p27 Q16, p32 Q49	A4
9.5 Quadratic expansion	p32 Q49	A4
9.7 Changing the subject of a formula	p26 Q6 , p31 Q40, p35 Q56	A5
10 Ratio, proportion and rates of change: Ratio, speed and proportion		
10.1 Ratio	p41 Q2 , Q6 , Q7 , p43 Q16, Q19, p61 Q53, p70 Q13	R4, R5, R6, R7
10.2 Speed, distance and time	p16 Q40, p28 Q27, p29 Q30, p46 Q29	R11
10.3 Direct proportion problems	p33 Q54, p41 Q1 , p42 Q10, Q12, p44 Q23	R10
10.4 Best buys	p16 Q44, p43 Q14, Q15, Q17, p44 Q20–Q22	R11, R12

Foundation student book chapter and section	Question	Curriculum ref.
11 Geometry and measures: Perimeter and area		
11.1 Rectangles	p18 Q56, p27 Q13, p42 Q12, p46 Q30, p53 Q10, p58 Q35, p59 Q38, Q40, p60 Q41	G16
11.2 Compound shapes	p53 Q9, p54 Q15, p55 Q19, Q20, p56 Q21, p59 Q39	G16
11.3 Area of a triangle	p55 Q19, p58 Q35, p59 Q39, p60 Q42	G16
11.4 Area of a parallelogram	p28 Q26	G16
11.5 Area of a trapezium	p55 Q19	G16
11.6 Circles	p62 Q63	G17
11.7 The area of a circle	p54 Q15, p56 Q24, p58 Q31	G17
11.8 Answers in terms of π	p56 Q24	G17
12 Geometry and measures: Transformations		
12.2 Translation	p53 Q11, p57 Q30, p62 Q60, p63 Q74	G7, G24
12.3 Reflections	p52 Q2 – Q4, p53 Q11, p61 Q58, p62 Q60	G7
12.4 Rotations	p52 Q1, Q3, p53 Q11, p60 Q43, p61 Q58, p62 Q60	G7
12.5 Enlargements	p52 Q5, p57 Q30, p62 Q59	G7
12.6 Using more than one transformation	p52 Q3, Q4, p57 Q30	G7, G8
12.7 Vectors	p57 Q30, p63 Q73, Q74	G24, G25
13 Probability: Probability and events		
13.1 Calculating probabilities	p69 Q1 – Q3, Q5, Q6, Q10, Q11, p70 Q13, Q14, Q17, p71 Q20	P2, P3, P4
13.2 Probability that an outcome will not happen	p70 Q17	P2, P4
13.3 Mutually exclusive and exhaustive outcomes	p69 Q7, p71 Q20, Q21	P2, P4
13.4 Experimental probability	p69 Q4, Q8, Q9, Q12, p70 Q18	P1, P2, P4, P5
13.5 Expectation	p69 Q4, p70 Q15, Q16, Q19, p71 Q23	P3, P4
13.6 Choices and outcomes	p16 Q46, p30 Q37	P4
14 Geometry and measures: Volumes and surface areas of prisms		
14.1 3D shapes	p62 Q67	N12, G16
14.2 Volume and surface area of a cuboid	p18 Q53, p19 Q61, p54 Q16, p59 Q40, p60 Q44, Q45, p62 Q66, Q68	G16
14.3 Volume and surface area of a prism	p62 Q61	G16, G17
14.4 Volume and surface area of cylinders	p46 Q31, p56 Q22, p58 Q34	G16, G17

Foundation student book chapter and section	Question	Curriculum ref.
15 Algebra: Linear equations		
15.1 Solving linear equations	p25 Q3 , p27 Q20 , Q21 , p31 Q43 , p31 Q45 , p32 Q48 , p35 Q56	A17
15.2 Solving equations with brackets	p33 Q51 , p58 Q35	A17
15.3 Solving equations with the variable on both sides	p33 Q51 , p58 Q35	A17
16 Ratio, proportion and rates of change: Percentages and compound measures		
16.1 Equivalent percentages, fractions and decimals	p33 Q52 , p41 Q3 , Q4 , Q8 , Q9 , p70 Q14	N10
16.2 Calculating a percentage of a quantity	p12 Q18 , p13 Q21 , p41 Q5 , Q8	N12, R9
16.3 Increasing and decreasing quantities by a percentage	p11 Q1 , p42 Q11 , p43 Q13 , Q18 , p45 Q24 , Q26 , p60 Q44	R9
16.4 Expressing one quantity as a percentage of another	p44 Q23 , p80 Q23	R9
16.5 Compound measures	p16 Q40	N13, R11
17 Ratio, proportion and rates of change: Percentages and variation		
17.1 Compound interest and repeated percentage change	p46 Q33 , Q34	R9, R16
17.2 Reverse percentage (working out the original value)	p45 Q25 , Q27	R9, R16
17.3 Direct proportion	p33 Q54 , p46 Q28	R7, R10, R13, R14
17.4 Inverse proportion	p33 Q54	R7, R10, R13, R14
18 Statistics: Representation and interpretation		
18.1 Sampling	p78 Q11 , p80 Q19 , p83 Q31 , Q32–Q34 , p85 Q45	S1
18.2 Pie charts	p76 Q2 , Q5 , p77 Q9 , p78 Q13 , p80 Q19 , p82 Q27 , p84 Q37	S2
18.3 Scatter diagrams	p80 Q22 , p82 Q26	S6
18.4 Grouped data and averages	p80 Q18 , Q23 , p84 Q36 , p85 Q42 , Q43	S4
19 Geometry and Measures: Constructions and loci		
19.1 Constructing triangles	p57 Q25 , p61 Q46 , Q47 , Q55	G1, G2
19.2 Bisectors	p57 Q25	G1, G2
19.3 Defining a locus	p61 Q48	G1, G2
19.4 Loci problems	p61 Q48	G1, G2

Foundation student book chapter and section	Question	Curriculum ref.
21 Algebra: Number and sequences		
21.1 Patterns in number	p28 Q28 , p29 Q33, p33 Q52	A23
21.2 Number sequences	p28 Q25, Q28, p30 Q35	A23–25
21.3 Finding the *n*th term of a linear sequence	p28 Q25, Q28, p30 Q35, p32 Q47	A23–25
21.4 Special sequences	p33 Q50	A23–25
21.5 General rules from given patterns	p30 Q34, p32 Q46	A23–25
22 Geometry and measures: Right-angled triangles		
22.1 Pythagoras' theorem	p53 Q10, p58 Q31, Q36	G6, 20
22.2 Calculating the length of a shorter side	p57 Q29, p58 Q32, p63 Q72	G6, 20
22.3 Applying Pythagoras' theorem in real-life situations	p57 Q29, p58 Q33, Q34, Q36, p61 Q57, p62 Q64, Q69, p63 Q70	G6, 20
22.5 Trigonometric ratios	p56 Q23	R12, G20, G21
22.6 Calculating lengths using trigonometry	p56 Q23, p63 Q71	G20, G21
22.7 Calculating angles using trigonometry	p56 Q23, p63 Q71	G20, G21
22.9 Solving problems using trigonometry	p56 Q23, p61 Q57, p62 Q65	G20, G21
23 Geometry and measures: Congruency and similarity		
23.1 Congruent triangles	p59 Q37	R12, G6, G7, G19
23.2 Similarity	p46 Q30 –Q32, p62 Q68	R12, G6, G7, G19
24 Probability: Combined events		
24.1 Combined events	p71 Q22, Q23, p72 Q24, Q26, Q27	P6, P7
24.2 Two-way tables	p71 Q23, p80 Q23	P6
24.3 Probability and Venn diagrams	p70 Q17	P6, P8
24.4 Tree diagrams	p71 Q21, p72 Q25, Q27	P1, P6, P8
25 Number: Powers and standard form		
25.1 Powers (indices)	p14 Q29 –Q32, p17 Q48, Q49, p18 Q53, p31 Q38	N6
25.2 Rules for multiplying and dividing powers	p14 Q30 –Q32, p17 Q49, p30 Q34	N6, A4
25.3 Standard form	p17 Q47, Q50	N9
27 Algebra: Non-linear graphs		
27.1 Distance–time graphs	p28 Q27	A14
27.2 Plotting quadratic graphs	p33 Q54	A18

basic

medium

hard

1 Number

1. **a** It takes 1.75 m of cotton to make a skirt. The cotton costs £3.50 per metre.
   How much will the material for the skirt cost?

   b The selling price of the skirt is double the cost of the material because of trimmings, labour and profit. How much will the maker charge for the skirt?

   c The price of cotton rises by 5%. How much does the material cost now?

2. A carton contains 1 litre of milk to the nearest 10 ml. What is the least amount of milk likely to be in the carton? Give your answer in millilitres.

3. Kilo means 1000, so 1 km = 1000 m.

   Centi means $\frac{1}{100}$th, so 1 cm = $\frac{1}{100}$th of a metre.

   Milli means $\frac{1}{1000}$th, so 1 mm = $\frac{1}{1000}$th of a metre.

   Use this information to find the units: 1 km = 100 000 __. Show your working.

4. Alex converted $1\frac{1}{4}$ hours to minutes. He worked it out as 125 minutes.

   What mistake did he make?

5. John and Peter are going to the cinema. The film starts at 18:40. It is $1\frac{3}{4}$ hours long plus 15 minutes for the trailers and adverts. Peter's dad is going to pick them up. He lives 20 minutes away by car.

   > If I leave at 9 pm I should get there just as they leave.

   Is Peter's dad correct? Explain your answer.

6. Look at the three fractions. Give a reason to justify why each one could be the odd one out.
   $\frac{1}{4}, \frac{2}{6}, \frac{2}{3}$

7. **a** Explain how you order numbers with decimals. Which part of the number do you look at to help you?

   b Write a number between 0.23 and 0.27. Which of the two numbers is it closest to? How do you know?

8. **a** Pens cost 85p. Krishnan has £5. He buys five pens. How much change does he get?

   b Which calculation is correct for answering this problem?
   - $500 \times (85 \times 5)$
   - $500 + (85 \times 5)$
   - $500 - (85 \times 5)$
   - $500 \div (85 \times 5)$

   c Write a simple word problem for each of the other calculations.

9 Immediately before Davina was paid, her bank balance was −£123.67.

    a What does the minus sign tell her about her account?

    b Immediately after she was paid, her balance was £1428.62. How much was she paid?

10 **a** I have 72 on the display of my calculator. What single operation should I key in to change it to 7200? Explain why it works.

    b The answer is 25. What could the question be? Write two different questions that have the same answer.

    c Write a different question that has an answer 10 times bigger than part **b**. Explain how you did this.

11 Annabel says 36 is a prime number and has eight factors. Explain why both statements cannot be true. Which one is true?

12 **a** Can a prime number be a multiple of 4? Explain your answer.

    b Give an example of a number greater than 500 that is divisible by 3. How do you know?

    c How do you know if a number is divisible by 6?

    d Give an example of a number greater that 100 that is divisible by 5 and also by 3. How do you know?

    e Is there a quick way to check if a number is divisible by 25? If so, what is it?

13 Sasha is fitting square floor tiles on a square kitchen floor.

Altogether she needs 16 tiles.

How many tiles are needed for each row?

14 Sarah divides a number by 10, and then again by 10. The answer is 7.2. What number did she start with? How do you know?

15 Asha says, 'I have 34p in 20p, 5p and 2p coins.' What is the minimum number of coins Asha could have?

16 **a** Which is longer: 300 cm or 30 000 mm? Explain how you worked it out.

    b Write another length that is the same as 5 m.

17 A shopkeeper buys 12 boxes of ice creams. Each box contains 15 ice creams and costs £4.50. He sells 105 of the ice creams for 80p each. He then sells the remaining ice creams for 50p each. What was his profit on the ice creams? Show all your working.

18 A mobile phone contract has a monthly fee of £25. This includes 300 minutes of free calls. Additional calls cost 5p per minute.

    a Calculate the cost of using the phone for 540 minutes in one month.

    b What difference would it make to the bill if calls after 6 pm cost 3p per minute and 25% of the calls were made after 6 pm?

19 **a** Give a fraction between $\frac{1}{3}$ and $\frac{1}{2}$, explaining how you worked it out.

    b How would you find out which of these fractions is closest to $\frac{1}{3}$?

$$\frac{10}{31}, \frac{20}{61}, \frac{30}{91}, \frac{50}{151}$$

20 What do you think is the main mistake students make when answering Q19?

21

a Mick wants to create a patio in his back garden. The diagram shows a plan of the area he wants to fill. He needs 78 bricks per square metre. How many bricks will he need?

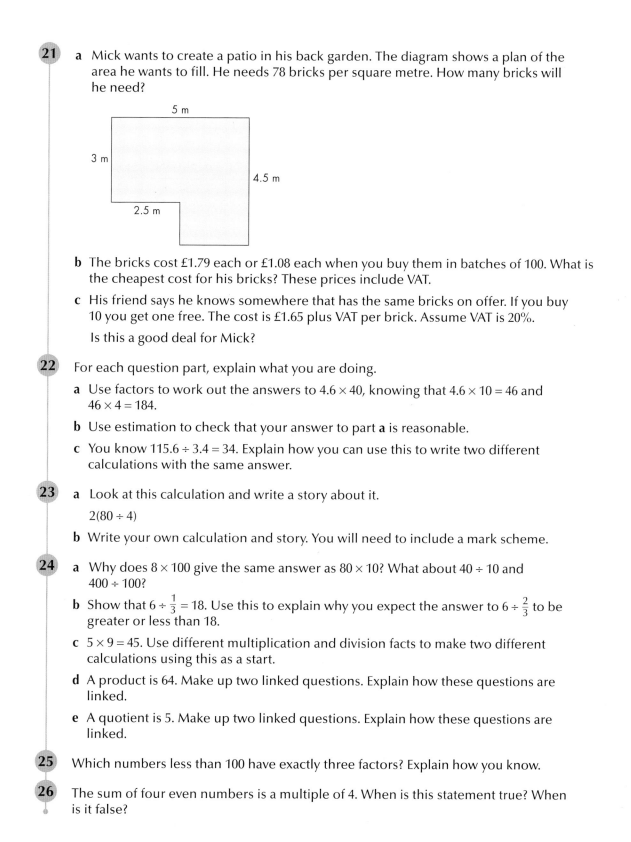

5 m

3 m

4.5 m

2.5 m

b The bricks cost £1.79 each or £1.08 each when you buy them in batches of 100. What is the cheapest cost for his bricks? These prices include VAT.

c His friend says he knows somewhere that has the same bricks on offer. If you buy 10 you get one free. The cost is £1.65 plus VAT per brick. Assume VAT is 20%.

Is this a good deal for Mick?

22 For each question part, explain what you are doing.

a Use factors to work out the answers to 4.6×40, knowing that $4.6 \times 10 = 46$ and $46 \times 4 = 184$.

b Use estimation to check that your answer to part **a** is reasonable.

c You know $115.6 \div 3.4 = 34$. Explain how you can use this to write two different calculations with the same answer.

23

a Look at this calculation and write a story about it.

$2(80 \div 4)$

b Write your own calculation and story. You will need to include a mark scheme.

24

a Why does 8×100 give the same answer as 80×10? What about $40 \div 10$ and $400 \div 100$?

b Show that $6 \div \frac{1}{3} = 18$. Use this to explain why you expect the answer to $6 \div \frac{2}{3}$ to be greater or less than 18.

c $5 \times 9 = 45$. Use different multiplication and division facts to make two different calculations using this as a start.

d A product is 64. Make up two linked questions. Explain how these questions are linked.

e A quotient is 5. Make up two linked questions. Explain how these questions are linked.

25 Which numbers less than 100 have exactly three factors? Explain how you know.

26 The sum of four even numbers is a multiple of 4. When is this statement true? When is it false?

27 **a** Write down:

      i a prime number greater than 100

      ii the largest cube number smaller than 1000

      iii two prime numbers that add up to 98.

  Show your working.

  b Give reasons why none of these numbers are prime numbers:

    2484, 17625, 3426

  c 629 is a product of two prime numbers. One of the numbers is 17. What is the other one?

28 Toby chooses three different whole numbers between one and 50.

The first number is a prime number.

The second number is three times the first number.

The third number is five more than the second number.

The sum of the numbers is greater than 25 but less than 40.

How many possible first numbers are there? What are they?

29 Are the following statements always, sometimes or never true? Justify your answers.

  a Cubing a number makes it bigger.

  b The square of a number is always positive.

30 Write each number as a power of a different number. The first one has been done for you.

  a $16 = 2^4$

  b $64 =$

  c $27 =$

  d $36 =$

31 In a warehouse, books are stored on shelves in piles of 20. There are 10 piles of books on each shelf.

  a How many books are on two shelves? Here is one way of writing the calculation. Fill in the gaps.

    $20 \times \underline{\ \ } \times \underline{\ \ } = 20 \times 20 = 20\underline{\ \ }$

  b Ben has just answered this question and he says there are $2^4 \times 5^5$ books on each shelf.

    How did he work this out?

  c Ben also says that if there are always an even number of shelves, the number of books on N pairs of shelves must be $2^3 \times 5^2 \times 2^N$.

    Explain why Ben is wrong.

32 **a** Write down the value of $36 \div 36$.

  b Write $6^2 \div 6^2$ as a single power of 6.

  c Use parts **a** and **b** to write down the value 6^0. What happens when you divide a number by the same number?

  d Write down one pair of possible values for a and b.

    $8^a \times 8^b = 8^9$

33 **a** Is each statement sometimes, always or never true? Justify your answers.

    i The more digits a number has, the bigger the number.

    ii Adding two numbers together gives a number that is bigger than either of the two original numbers.

    iii Calculating a fraction of a number makes it smaller.

  b Give two fractions that multiply together to give a bigger answer than either of the fractions you are multiplying.

  c Explain why the fractions you chose work.

  d Is this statement true or false? Explain your answer.

    10 is greater than 9, so 0.10 must be greater than 0.9.

34 **a** The lowest winter temperature in a city in Norway was −15 °C. The highest summer temperature was 42 °C higher. What was the highest summer temperature?

  b The answer is to a question −7. Make up two addition/subtraction calculations with the same answer. Make one easier and one harder.

35 Is the answer to 68 ÷ 0.8 smaller or larger than 68? Explain your answer.

36 **a** Rewrite this paragraph using sensible numbers:

    I left home at eleven and a half minutes past two, and walked for 49 minutes.

    The temperature was 12.7623 °C. I could see an aeroplane overhead at 2937.1 feet.

    Altogether I walked 3.126 miles.

  b Write your own question similar to the one in part **a**, together with a mark scheme.

37 Sarah has an old map where the scale is in miles. 1 mile is approximately equal to $\frac{8}{5}$ km. A new supermarket has opened in a nearby town which is shown as $5\frac{1}{2}$ miles on this map. Her usual supermarket is 7 km away. She wants to know which one is closer. Show your working.

38 Mrs Smith pays her electricity bill monthly by standing order. To make sure she is paying enough to cover her bill, she has read her meter. At the start of the month it read 78 987 kWh. At the end of the month it read 79 298 kWh.

Mrs Smith pays 20.95 pence per kWh for the first 80 kWh she uses and 10.8 pence per kWh for every other kWh used.

She pays £50 on her standing order. Will this cover her bill? Show your working and justify any assumptions you make.

39 Mr Jones is considering changing his electricity supplier. With his current supplier he pays 15.09 pence per kWh. The new supplier charges 14.37 pence per kWh.

In addition, the new supplier will charge a fixed fee of 23.818 pence per day but his current supplier only charges him 13.99 pence per day.

The table shows how much electricity he used last year.

Should Mr Smith switch suppliers? Explain your answer fully.

Period	Previous reading	Current reading
27 August–30 December	53 480	54 782
31 December–9 April	54 782	55 916

40 **a** For each example, write down the basic calculation you need for the question. Then describe how you will adapt the answer to suit each context.

      i At a wedding, there are 175 guests sitting at tables of 8. How many tables are needed?

      ii When 8 colleagues go out for lunch they agree to split the bill of £175 equally. How much should each person pay?

      iii Jessy has 175 bread rolls and is packing them in boxes of 8. How many boxes can she fill and how many bread rolls will she have left over?

      iv Ben travels 175 kilometres in 8 hours. What is his average speed in kilometres per hour?

  b Design a similar set of questions and a mark scheme of your own.

41 **a** $24 \times 72 = 1728$. Explain how you can use this fact to write calculations with answers 17.28.

  b Design your own question like the one in part **a**.

42 A parking space is 4.8 metres long to the nearest tenth of a metre.

A car is 4.5 metres long to the nearest half metre.

Which of the following statements is definitely true?

A: The space is big enough.

B: The space is not big enough.

C: It is impossible to tell whether or not the space is big enough.

Explain how you decide.

43 Billy has 20 identical bricks. Eack brick is 15 cm long, to the nearest centimetre.

The bricks are put end to end to build a wall. What is the greatest possible length of the wall?

44 Barry has a four-wheel drive vehicle that uses a lot of petrol. It will cost him £2500 to convert the car so it can use a cheaper fuel.

  a Write a question for Barry to answer to help him decide if he is going to convert his car.

  b What assumptions will Barry need to make and what pieces of information will Barry need to answer his question?

  c Using your answer to parts **a** and **b**, present an argument to help Barry decide if he should get his car converted.

45 A dolphin has to breathe even when it is asleep in the water. Alison is a marine biologist who studies dolphins.

She observes that a dolphin takes a breath at the surface, dives to the bottom of the sea and starts to sleep.

From the bottom it floats slowly to the surface in five minutes and then takes another breath. After another two minutes it is back at the bottom of the sea.

Alison watches the dolphin for 1.5 hours. At the end of the 1.5 hours is the dolphin at the bottom, on its way up, at the top or on its way down? Explain your answer.

46 How many ways can you label a product if the label must consist of a letter AND a number from 1 to 25.

47 **a** Copy the table and rewrite the distance and the diameter in standard form.

Planet	Distance from the sun (million km)	Mass (kg)	Diameter (km)
Mercury	58	3.3×10^{23}	4878
Venus	108	4.87×10^{24}	12104
Earth	150	5.98×10^{24}	12756
Mars	228	6.42×10^{23}	6787
Jupiter	778	1.90×10^{27}	142796
Saturn	1427	5.69×10^{26}	120660
Uranus	2871	8.68×10^{25}	51118
Neptune	4497	1.02×10^{26}	48600

b i Which planet is the largest?

  ii Which planet is the smallest?

  iii Which planet is the lightest?

  iv Which planet is the heaviest?

  v Which planet is approximately twice as far away from the sun as Saturn?

  vi Which two planets are similar in size?

48 Are these statements always, sometimes or never true? If a statement is sometimes true, when is the statement true and when is it false?

  a You can work out the square root of any number.

  b You can work out the cube root of any number.

49 **a** Show that:

  i $5^6 \div 5^{-3} = 5^9$

  ii $5^6 \times 5^{-3} = 5^3$

  b What does the index of $\frac{1}{2}$ represent?

50 The diameter of the smallest virus is 20 nanometres. The height of the tallest skyscraper in 2015 is the Burj Khalifa in Dubai. It is 8.298×10^2 m high. The height of Mount Everest is 8.848×10^3 m.

  a How many times taller is the mountain than the skyscraper?

  b How high is the skyscraper in km?

  c A nanometre is $\frac{1}{1\,000\,000\,000}$ th of a meter. Write the diameter of the virus in metres in standard form.

51 The diagram shows three identical shapes A, B and C.

$\frac{5}{7}$ of shape A is shaded.

$\frac{8}{9}$ of shape C is shaded.

What fraction of shape B is shaded?

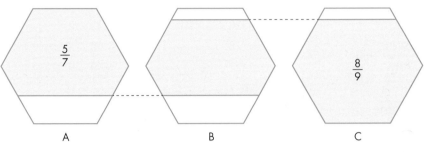

A   B   C

52 In a city, $5\frac{1}{2}$ out of every 15 square metres is used for residential development and services. $\frac{5}{8}$ of the residential land is actually used for housing. What percentage of the total area of the city is used for services?

53 Siobhan runs a company which manufactures small plastic items. One item the company makes is a solid plastic dice with a side of 2 cm.

Siobhan wants to make a larger dice. She needs to know what it will cost. This will depend on the amount of plastic required to make it.

She thinks that a dice with a side of 4 cm will require twice as much plastic to make as a dice with a side of 2 cm. Explain why this is not the case. How much more plastic will be required?

54 Estimate an answer to each calculation. Is your estimate higher or lower that the real answer? Explain why this is.

a 75.8×23.1

b $\frac{26.2}{3.5}$

55 Give three examples of multiplication and division calculations, with answers that approximate to 75. Try to make each example harder than the last. Explain how you decided on each of the calculations.

56 The dimensions of a classroom floor, measured to the nearest metre, are 19 m by 15 m. What range must the area of the floor lie within? Suggest a sensible answer for the area, given the degree of accuracy of the measurements.

57 You are the manager of a transport company. You own a lorry with six axles and a lorry with five axles. The five-axle lorry is 10% cheaper to run per trip than the six-axle lorry.

These are the load restrictions for heavy goods vehicles.

Type of lorry	Load limit
six axles	44 tonnes
five axles	40 tonnes

You deliver pallets of goods.

Each pallet weighs 500 kg to the nearest 50 kg. Explain which lorry would you use for each job, showing clearly any decisions you make.

a A company orders 80 pallets.

b A company orders 150 pallets.

c A company orders 159 pallets.

58 **a** Explain the difference in meaning between 0.4 m and 0.400 m. Why is it sometimes necessary to include the zeros in measurements?

b What range of measured lengths might be represented by the measurement 430 cm?

c What accuracy is needed to be sure a measurement is accurate to the nearest metre?

d Why might you pick a runner for a race whose time for running 100 m is recorded as 13.3 seconds rather than 13.30 seconds? Why might you not?

e Explain how seven people, each with a mass of 100 kg, might exceed a limit of 700 kg for a lift.

59 A theatre has 365 seats.

For a show, 280 tickets are sold in advance.

The theatre's manager estimates that another 100 people to the nearest 10 will turn up without tickets.

Is it possible they will all get a seat, assuming that 5% of those with tickets do not turn up?

Show clearly how you decide.

60 A stopwatch records the time for a winner of a 100-metre race at 12.3 seconds, measured to the nearest one-tenth of a second. The length of the track is correct to the nearest centimetre. What is the fastest possible average speed of the winner?

61 A cube has a volume of 125 cm³ to the nearest 1 cm³. Work out the limits of accuracy of the area of one side of the square base.

⑦Hints and tips

Question	Hint
2	1 litre = 1000 ml
6	You could compare them to fractions you know well or decide if they are multiples of each other.
40	Think about what might be a sensible answer. Can you have part of a table or a fraction of a penny?

Worked exemplars

 1 I want to buy a burger, a portion of chips and a bottle of cola for my lunch. I have the following coins in my pocket: £1, 50p, 50p, 20p, 20p, 10p, 2p, 2p, 1p.

Price List	
Burger	£1.20
Chips	90p
Beans	50p
Cola	60p

a Do I have enough money?

b Do I have enough money if I replace the chips with beans?

This is a question for which you need to use and show your mathematical reasoning skills, so set out your answer clearly.	
a Total cost for a burger, chips and a cola = £2.70 Total of coins in my pocket = £2.55	Show the total cost of the meal and the total of the coins in your pocket.
No, as £2.70 > £2.55	It is important to show a clear conclusion. Do not just say 'No'. Use the numbers to support your answer.
b Beans are 40p cheaper and £2.30 < £2.55, so, yes I have enough money.	Explain clearly why you have enough money for this meal.

2 A PE teacher is organising 20 students into equal-sized teams for a competition. How many can she have in each team?

This is a mathematical reasoning question, so you need to construct a chain of reasoning to reach the required result.	
Factors of 20 are {1, 2, 4, 5, 10, 20}. 1 student alone would not make a team. 20 students in a team could not play in a competition. So the teacher can have teams of 2, 4, 5 or 10.	The teams will need to be of equal size and there will have to be more than one student in a team so they can play each other. So, although 1 and 20 are factors, they cannot form part of the final answer as this does not address the team requirement.

3 Give a reason why each of these numbers could be the odd one out.

123   144   169

This is a 'communicating mathematics' question, so make sure your answers are clear and that you use correct mathematical terms.	
123 because it is not a square number. 144 because it is not odd. 169 because it is not a multiple of 3.	Answer the whole question by giving a reason for each number being the odd one out and not just identifying one of the numbers as the odd one out. Each answer is for any valid reason, so there could be other solutions.

4 The perimeter of a square is 36 cm.

This length is accurate to the nearest centimetre.

Work out the error interval due to rounding of the length of a side of the square.

This is a problem-solving question, so you need to make connections between different topics in mathematics (in this case properties of squares and rounding) and to show your strategy clearly.	
Perimeter = 4 × length of one side	Write down the correct formula for the perimeter of a square.
Error interval of perimeter is: $35.5 \leqslant$ perimeter < 36.5	Write down the error interval of the perimeter.
Error interval of one side is $\frac{1}{4}$ of the error interval of the perimeter.	Show that the error interval of a side is a quarter of the error interval of the perimeter.
Error interval of one side is: $8.875 \leqslant$ length of side < 9.125	Write down the error interval of the side.

5 Use estimation to put the expressions below in order of size, starting with the smallest.

    A $\dfrac{9.7 \times 10.3}{7.2 - 2.1}$ **B** $88.8 \div 5.8$ **C** $\left(\dfrac{3.7 + 6.2}{1.9}\right)^2$

This is a mathematical reasoning question. You need to show your working and your conclusion.	
A: $\dfrac{10 \times 10}{7 - 2} = \dfrac{100}{5} = 20$	Round all the numbers in each expression to 1 sf and work out the approximate value.
B: $90 \div 6 = 15$	
C: $\left(\dfrac{4 + 6}{2}\right)^2 = \left(\dfrac{10}{2}\right)^2 = 5^2 = 25$	
The order from smallest is: B, A, C	Write down the order, starting with the smallest.

6 The length of a piece of wood is measured as 2.3 metres to the nearest 10 cm. Which of these is the error interval for the length?

    A   2.2 m $\leqslant$ length < 2.3 m   **B**   1.3 m $\leqslant$ length < 3.3 m

    C   2.25 m $\leqslant$ length < 2.35 m   **D**   2.29 m $\leqslant$ length < 2.31 m

This is a mathematical reasoning question as there is a mixture of units. The length is given in metres but the accuracy is measured in centimetres.	
2.3 metres = 230 centimetres	Change the length to centimetres.
Error interval is $225 \leqslant$ length < 235	An accuracy of 10 cm means that the length is within ± 5 cm.
2.25 m $\leqslant$ length < 2.35 m which is choice C.	Convert the error interval back to metres.

7 The perimeter of a rectangle is $32\frac{1}{2}$ cm.

Use your calculator to work out one pair of possible values for the length and the width of the rectangle.

This is a problem-solving question, so you need to make connections between different part of mathematics (in this case perimeters, substitution and fractions) and show your strategy clearly.	
Perimeter = 2 × length + 2 × width	Write down the correct formula for the perimeter of a rectangle.
2 × length + 2 × width = $32\frac{1}{2}$ length + width = $32\frac{1}{2} \div 2$	Put the value that you know into the formula.
length + width = $16\frac{1}{4}$ cm	Work out the total value of the length and width.
Possible length and width are: Length = 10 cm Width = $6\frac{1}{4}$ cm	Any two values with a sum of $16\frac{1}{4}$ would answer the question. Choosing 10 makes the calculation straightforward.

8 Bob needs to travel 187 miles to a business meeting.

He sets off at 10:00 am and wants to arrive before the meeting starts at 3:00 pm.

He stops for one 20-minute break and his average speed when driving is $42\frac{1}{2}$ mph.

Will he make it on time?

This is a mathematical reasoning question. There is a lot of information given, so make sure you show a clear chain of reasoning to get to your result.	
Time travelling = $187 \div 42\frac{1}{2} = 4\frac{2}{5}$ hours	Work out the time it takes to drive the distance.
$4\frac{2}{5}$ hours = 4 hours 24 minutes	Change the fraction into hours and minutes. $\frac{1}{5}$ of an hour is 12 minutes
Total time = 4 hours 24 minutes + 20 minutes = 4 hours 44 minutes	Add on the time taken for a break.
10:00 am + 4 hours 44 minutes = 2:44 pm, so he will arrive on time.	Add the total time to 10:00 am and write a conclusion.

9 a Which of these is *not* equivalent to $3 \times 10^2 \times 6 \times 10^3$?

18×10^5 1.8×10^6 180 000   1 800 000

b Arrange these expressions in order of size, starting with the smallest.

6.3×10^3 64×10^2 6250 $130\,000 \div 20$

This is a problem-solving question, so you need to show a clear strategy in your solution, and show all the steps in your answer.

a $3 \times 10^2 \times 6 \times 10^3$	Work out the value of $3 \times 10^2 \times 6 \times 10^3$ and then compare it to the values given.
$= (3 \times 6) \times (10^2 \times 10^3)$	
$= 18 \times 10^5$	
$= 1\,800\,000$	
$= 1.8 \times 10^6$	
So, 180 000 is not equivalent to $3 \times 10^2 \times 6 \times 10^3$.	
b $6.3 \times 10^3 = 6300$	Work out the value of each expression in the same format (either an ordinary number or standard form). The ordinary numbers do not have that many digits, so this is the more straightforward format in this case.
$64 \times 10^2 = 6400$	
6250	
$130\,000 \div 20 = 13\,000 \div 2$	
$= 6500$	
So, in order:	
$6250,\ 6.3 \times 10^3,\ 64 \times 10^2,\ 130\,000 \div 20$	

---

10 This is a table of powers of 3.

3^1	3^2	3^3	3^4	3^5	3^6	3^7
3	9	27	81	243	729	2187

a Use your calculator to work out $27 \div 243$. Give the answer as a fraction.

b Use the rules of indices to write $3^3 \div 3^5$ as a single power of 3.

c Deduce the value, as a fraction, of 3^{-3}.

This is a mathematical reasoning question. The first two parts set up the information you will need. Make sure that you show your steps and use connecting words or symbols such as $\Rightarrow$ (implies) and $\therefore$ (therefore).

a $27 \div 243 = \frac{1}{9}$	Enter $27 \div 243$ as a fraction. This should cancel to the simplest form. Make sure you know how to change an answer into a fraction if the display shows a decimal, in this case 0.111…
b $3^3 \div 3^5 = 3^{(3-5)}$ $= 3^{-2}$	Apply the rules of indices. When dividing powers with the same base, subtract them.
c Since $3^3 = 27$ and $3^5 = 243$, parts **a** and **b** $\Rightarrow \frac{1}{9} = 3^{-2}$ $\therefore 3^{-3} = \frac{1}{27}$	This is the mathematical reasoning section. Parts **a** and **b** are linked in that they are the same calculation in different forms, so the answers must be the same. So if $\frac{1}{9} = 3^{-2}$, then 3^{-3} must be $\frac{1}{27}$.

2 Algebra

Solve $\frac{x}{3} + 1 = 5$

Subtract 1 from both sides: $\frac{x}{3} = 4$

Multiply both sides by 3: $x = 12$

Check the answer by substituting into the original equation: $\frac{12}{3} + 1 = 4 + 1$

$$= 5$$

Adam opened a packet of biscuits and ate two of them. He then shared the rest with four friends. Each friend received three biscuits.

How many biscuits were in the packet originally?

Set up the equation. If there were x biscuits, he took away two and then shared the rest ($x - 2$ biscuits) between five people.

So $\frac{x - 2}{5} = 3$

Multiply both sides by 5: $x - 2 = 15$

Add 2 to both sides: $x = 17$

There were 17 biscuits in the packet originally.

Check the answer by substituting into the original equation: $\frac{17 - 2}{5} = \frac{15}{5}$

$$= 3$$

1 Write a formula for each of these:

   a number of days = 7 × the number of weeks

   b cost = price of one item × number of items

   c age in years = age in months ÷ 12

   d pence = number of pounds × 100

   e area of rectangle = length × width

   f cost of petrol for a journey = cost per litre × number of litres used.

2 **a** Write a formula for:

     cost of a vet's bill = £80 per hour multiplied by the number of hours

   b How could you change your formula to include a £50 call out fee?

3 **a** I think of a number and add 15. Do you know what my number is? Explain your answer.

   b I think of a number and add 15. The answer is 26. Do you know what my number is? Explain your answer.

4 Explain why it is possible for more than one calculation to match the same rule.

5 We use conventions in mathematics so everyone understands what we have written.

What are the important conventions when describing a point using coordinates? Use an example to explain why it is important that everyone follows the same conventions.

6 a Could the equation $y = 2x + 6$ be expressed in a different way, but still be the same? What is the mathematical term we use to describe this process?

   b Is $\frac{y}{2} = x + 3$ the same as $y = 2x + 6$? Justify your answer.

7 How do you know the point (3, 6) is not on the line $y = x + 2$?

Real-life questions often have a lot of words. You need to decide which ones are important and how to translate them into mathematics.

Example 3

Nevaeh needs to get to the station to catch a train. Her local taxi firm charges 50p per mile plus a call out fee of £2. Nevaeh has £5 cash and the station is 10 miles away from where she lives. Does she have enough money?

First you need to decide exactly what the question is you need to answer. Underline the important information.

Nevaeh needs to get to the station to catch a train. Her local taxi firm charges <u>50p per mile</u> plus a <u>call out fee of £2</u>. Nevaeh has <u>£5 cash</u> and the station is <u>10 miles</u> away from where she lives. <u>Does she have enough money?</u>

So the question is: is £5 enough money to travel 10 miles at a cost of 50p per mile plus a call out fee of £2?

First write a formula:

$C = 0.5n + 2$

where C is the cost of the taxi and n is the distance travelled.

2 is a constant that describes the call out fee.

Note that 50p per mile has been converted to £0.50 per mile. All of the parts should be in the same units.

Substitute $n = 10$ into the formula:

$C = 0.5 \times 10 + 2$

$\quad = 5 + 2$

$\quad = £7$

So as Nevaeh only has £5 she does not have enough money to pay for the taxi.

Check that you have answered the question.

8 Isla goes shopping with her friend Savannah. She buys two 10 CDs at £14.99 each. She also buys two coffees. They cost £2.50 each.

Isla would like to take a taxi home, which is 12 miles away from the shopping centre. The taxi firm charges 80p per mile plus a call out fee of £2.50. Isla had £70 when she left home. Isla is not sure if she has enough money to pay for the taxi.

   a What is the question?

   b Write a formula to help you answer the question. Explain why your formula works. Then use it to answer the question in part **a**.

9 How do you link a formula expressed in words to a formula expressed using algebra? Give an example.

10 Explain the difference between each pair of expressions:

   a $2n$ and $n + 2$

   b $3(c + 5)$ and $3c + 5$

   c n^2 and $2n$

11 A rectangle's length is three times its width. Its perimeter is 48 cm. Work out its area.

12 Here is a number line.

Work out the value of C.

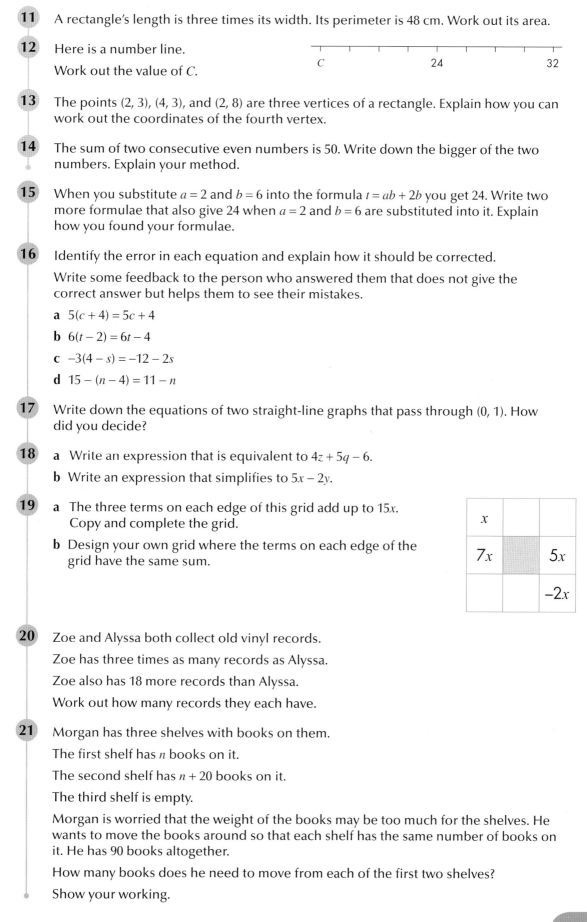

13 The points (2, 3), (4, 3), and (2, 8) are three vertices of a rectangle. Explain how you can work out the coordinates of the fourth vertex.

14 The sum of two consecutive even numbers is 50. Write down the bigger of the two numbers. Explain your method.

15 When you substitute $a = 2$ and $b = 6$ into the formula $t = ab + 2b$ you get 24. Write two more formulae that also give 24 when $a = 2$ and $b = 6$ are substituted into it. Explain how you found your formulae.

16 Identify the error in each equation and explain how it should be corrected.

Write some feedback to the person who answered them that does not give the correct answer but helps them to see their mistakes.

a $5(c + 4) = 5c + 4$

b $6(t - 2) = 6t - 4$

c $-3(4 - s) = -12 - 2s$

d $15 - (n - 4) = 11 - n$

17 Write down the equations of two straight-line graphs that pass through (0, 1). How did you decide?

18 **a** Write an expression that is equivalent to $4z + 5q - 6$.

b Write an expression that simplifies to $5x - 2y$.

19 **a** The three terms on each edge of this grid add up to 15x. Copy and complete the grid.

b Design your own grid where the terms on each edge of the grid have the same sum.

20 Zoe and Alyssa both collect old vinyl records.

Zoe has three times as many records as Alyssa.

Zoe also has 18 more records than Alyssa.

Work out how many records they each have.

21 Morgan has three shelves with books on them.

The first shelf has n books on it.

The second shelf has $n + 20$ books on it.

The third shelf is empty.

Morgan is worried that the weight of the books may be too much for the shelves. He wants to move the books around so that each shelf has the same number of books on it. He has 90 books altogether.

How many books does he need to move from each of the first two shelves?

Show your working.

22 How would you find coordinates for the straight-line graph $y = 2x + 3$ that are in the third quadrant?

23 **a** The points (–2, 3), (1, 3) and (5, 3) lie on a straight line. Work out the equation of the line.

   b The points (–2, –4), (1, 2), (3, 6) lie on another straight line. What is the equation of the line for this graph?

   c Work out three more points for the line in part **b** in the third quadrant without drawing the graph.

24 Explain how you know the point (3, 9) is not on the line $y = 2x + 2$.

25 Two sequences are:

5, 11, 17, 23, 29, 35, 41, ...

1, 4, 7, 10, 13, 16, 19, ...

Do the two sequences ever have any terms in common? Explain your answer.

26 The diagram shows two sides, AB and AC, of an isosceles trapezium ABCD.

   a Write down the coordinates of D.

   b Work out at the area of the trapezium ABCD.

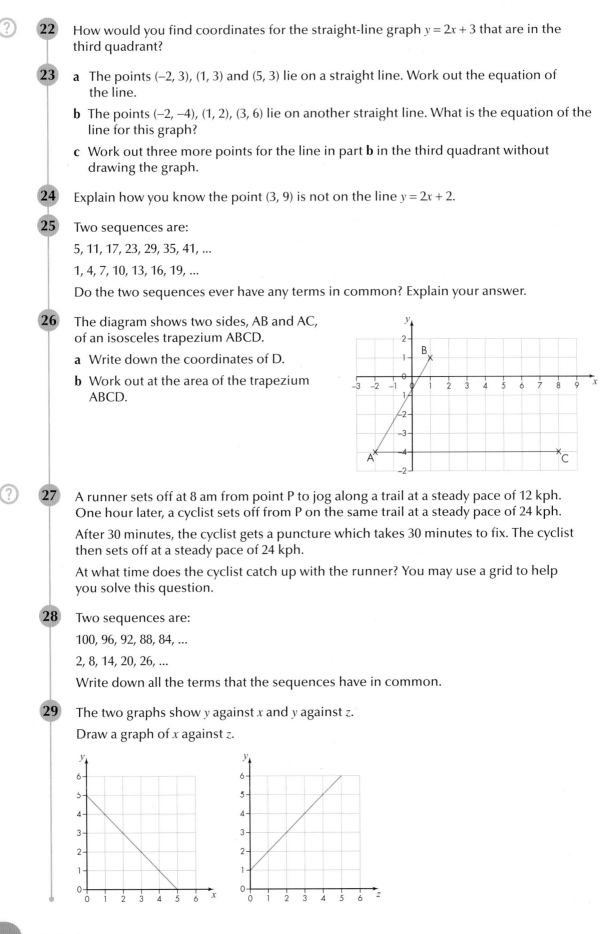

27 A runner sets off at 8 am from point P to jog along a trail at a steady pace of 12 kph. One hour later, a cyclist sets off from P on the same trail at a steady pace of 24 kph.

After 30 minutes, the cyclist gets a puncture which takes 30 minutes to fix. The cyclist then sets off at a steady pace of 24 kph.

At what time does the cyclist catch up with the runner? You may use a grid to help you solve this question.

28 Two sequences are:

100, 96, 92, 88, 84, ...

2, 8, 14, 20, 26, ...

Write down all the terms that the sequences have in common.

29 The two graphs show y against x and y against z.

Draw a graph of x against z.

30 Philip's journey is shown on the graph.

    a What is his average speed?

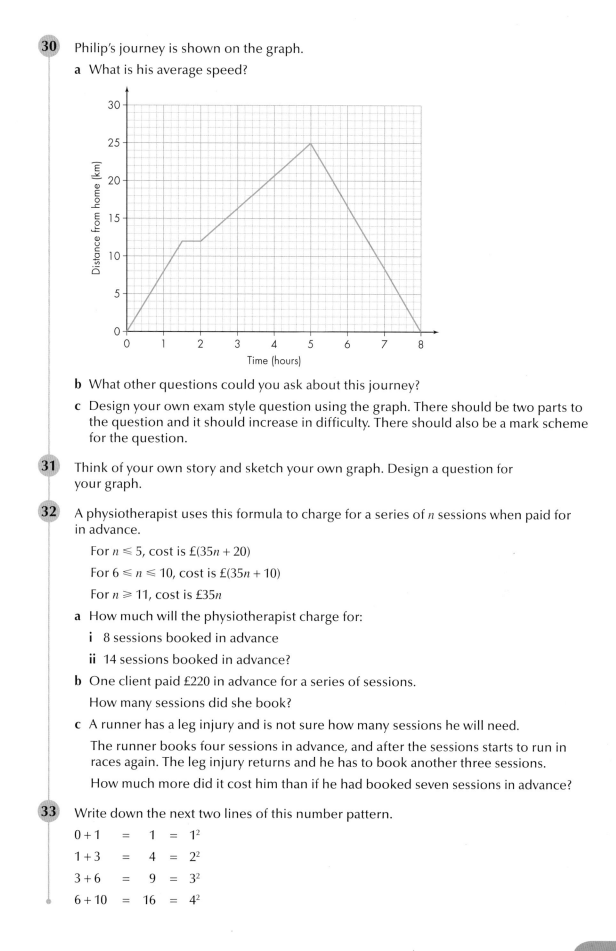

    b What other questions could you ask about this journey?

    c Design your own exam style question using the graph. There should be two parts to the question and it should increase in difficulty. There should also be a mark scheme for the question.

31 Think of your own story and sketch your own graph. Design a question for your graph.

32 A physiotherapist uses this formula to charge for a series of n sessions when paid for in advance.

    For $n \leq 5$, cost is £$(35n + 20)$

    For $6 \leq n \leq 10$, cost is £$(35n + 10)$

    For $n \geq 11$, cost is £$35n$

    a How much will the physiotherapist charge for:

      i 8 sessions booked in advance

      ii 14 sessions booked in advance?

    b One client paid £220 in advance for a series of sessions.

      How many sessions did she book?

    c A runner has a leg injury and is not sure how many sessions he will need.

      The runner books four sessions in advance, and after the sessions starts to run in races again. The leg injury returns and he has to book another three sessions.

      How much more did it cost him than if he had booked seven sessions in advance?

33 Write down the next two lines of this number pattern.

$0 + 1 \quad = \quad 1 \quad = \quad 1^2$

$1 + 3 \quad = \quad 4 \quad = \quad 2^2$

$3 + 6 \quad = \quad 9 \quad = \quad 3^2$

$6 + 10 \quad = \quad 16 \quad = \quad 4^2$

34 **a** Draw an equilateral triangle with each side 9 cm.

The perimeter will be 27 cm.

b Draw another equilateral triangle of side 3 cm on each edge.

Work out the perimeter of the new shape.

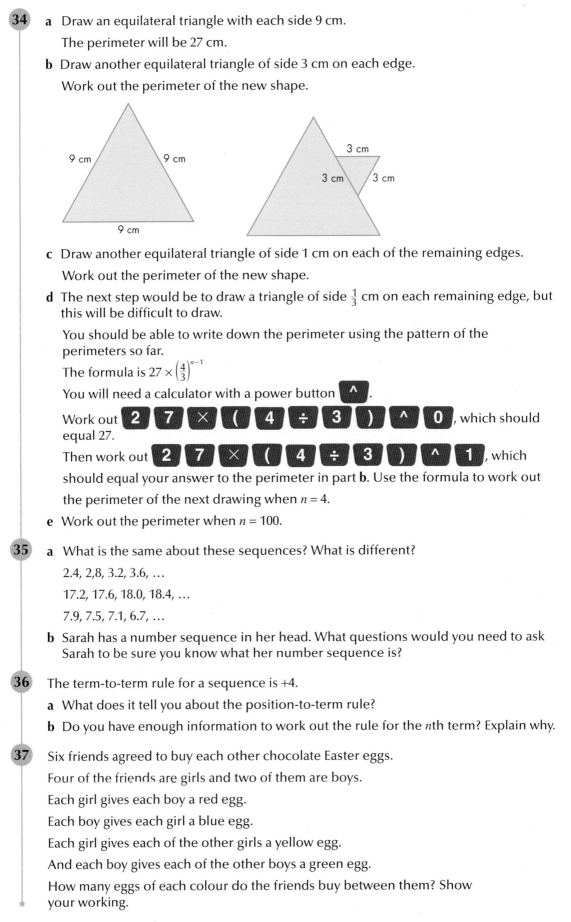

c Draw another equilateral triangle of side 1 cm on each of the remaining edges.

Work out the perimeter of the new shape.

d The next step would be to draw a triangle of side $\frac{1}{3}$ cm on each remaining edge, but this will be difficult to draw.

You should be able to write down the perimeter using the pattern of the perimeters so far.

The formula is $27 \times \left(\frac{4}{3}\right)^{n-1}$

You will need a calculator with a power button $\boxed{\wedge}$.

Work out $\boxed{2}\boxed{7}\boxed{\times}\boxed{(}\boxed{4}\boxed{\div}\boxed{3}\boxed{)}\boxed{\wedge}\boxed{0}$, which should equal 27.

Then work out $\boxed{2}\boxed{7}\boxed{\times}\boxed{(}\boxed{4}\boxed{\div}\boxed{3}\boxed{)}\boxed{\wedge}\boxed{1}$, which should equal your answer to the perimeter in part **b**. Use the formula to work out the perimeter of the next drawing when $n = 4$.

e Work out the perimeter when $n = 100$.

35 **a** What is the same about these sequences? What is different?

2.4, 2,8, 3.2, 3.6, …

17.2, 17.6, 18.0, 18.4, …

7.9, 7.5, 7.1, 6.7, …

b Sarah has a number sequence in her head. What questions would you need to ask Sarah to be sure you know what her number sequence is?

36 The term-to-term rule for a sequence is +4.

a What does it tell you about the position-to-term rule?

b Do you have enough information to work out the rule for the nth term? Explain why.

37 Six friends agreed to buy each other chocolate Easter eggs.

Four of the friends are girls and two of them are boys.

Each girl gives each boy a red egg.

Each boy gives each girl a blue egg.

Each girl gives each of the other girls a yellow egg.

And each boy gives each of the other boys a green egg.

How many eggs of each colour do the friends buy between them? Show your working.

(?) **38** Explain the difference between $2n^2$ and $(2n)^2$.

(?) **39** Explain how you know when a letter symbol represents an unknown or a variable.

40 Look at the list of formulae.

   i $V = lwh$

   ii $s = ut + \frac{1}{2}at^2$

   iii $x = 3y - 2$

   iv $s = \left(\dfrac{u + v}{2}\right)t$

   v $v^2 = u^2 + 2ad$

   vi $A = 2\pi r^2 + 2\pi rh$

  a Might it be difficult to substitute values into any of them? Which ones? Explain what makes them difficult and identify what the mistakes might be.

  b Might it be difficult to rearrange any of them? Which ones? Explain what makes them difficult and identify what the mistakes might be.

41 When you substitute $s = 6$ and $t = 2$ into the formula $z = \dfrac{3(s + 2t)}{12}$ you get 2.5.

Make up two more formulae that also give $z = 2.5$ when $s = 6$ and $t = 2$ are substituted.

Explain what you are doing and try to use two different methods as well.

An identity can be described as two expressions that mean exactly the same thing. Alternatively, it can be described as an equation that is TRUE whatever the value of the variable.

For example: $\frac{a}{4} = a \div 4$ or $2n(n + 4) = 2n^2 + 8m$ or $2^a \times 2^b = 2^{a + b}$

42 Write the expression $\dfrac{2n + 6}{2}$ in a different way.

43 The height of an isosceles triangle is three times its base. The area is 6 cm². What is its height?

44 **a** What method could you use to solve these problems?

    A number plus its cube is 20. What is the number?

    The length of a rectangle is 2 cm longer than the width. The area is 67.89 cm². What is the width?

  b Use this method to solve both problems.

45 **a** Explain why 'I think of a number and double it' is different from 'I think of a number and double it. The answer is 12.'

  b $6 = 2z - 8$

    i How many solutions does this equation have?

    ii Write another equation with the same solution.

    iii Why do they have the same solution? How do you know?

46 Harry is building three different patterns with counters. He builds the patterns in steps.

	Step 1	Step 2	Step 3	Step 4
Pattern 1	○ ○ ○	○ ○ ○ ○ ○ ○	○ ○ ○ ○ ○ ○ ○ ○ ○	○ ○ ○ ○ ○ ○ ○ ○ ○ ○ ○ ○
Pattern 2	○ ○	○ ○ ○ ○	○ ○ ○ ○ ○ ○	○ ○ ○ ○ ○ ○ ○ ○
Pattern 3	○	○ ○	○ ○ ○	○ ○ ○ ○

Harry has a packet that contains 1000 counters.

Which step will Harry get to before he runs out of counters?

47 Here are the first five terms in an arithmetic sequence.

3, 7, 11, 15, 19, …

Is 62 a term in this sequence? Explain your answer.

48 In the first round of voting for a school council Hussein got 10% more votes than Eliza.

Eliza got 50 votes fewer than Charles.

Denise got two-thirds of Eliza's votes.

Charles got 500 votes.

Each student needs a minimum of 350 votes to stay in the next round. How many contestants will there be in the next round? Show your working.

49 Use this diagram of an $(x + 1)$ by $(x + 1)$ square to show that $(x + 1)^2 = x^2 + 2x + 1$.

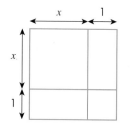

50 Pascal's triangle has many uses in mathematics.

Each row starts and ends with 1. The other numbers are formed by adding the two values above them.

a Continue Pascal's triangle for another three rows.

b Describe any patterns or special sequences you can see in Pascal's triangle.

c What is the special name given to the series of numbers down the diagonal marked B?

d Add each row, e.g. $1 = 1$, $1 + 1 = 2$, $1 + 2 + 1 = 4$.

e Explain how the series formed is building up.

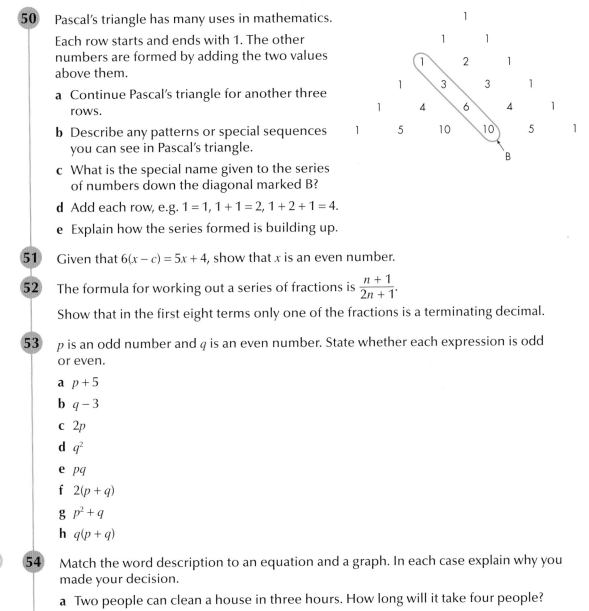

51 Given that $6(x - c) = 5x + 4$, show that x is an even number.

52 The formula for working out a series of fractions is $\dfrac{n + 1}{2n + 1}$.

Show that in the first eight terms only one of the fractions is a terminating decimal.

53 p is an odd number and q is an even number. State whether each expression is odd or even.

a $p + 5$

b $q - 3$

c $2p$

d q^2

e pq

f $2(p + q)$

g $p^2 + q$

h $q(p + q)$

54 Match the word description to an equation and a graph. In each case explain why you made your decision.

a Two people can clean a house in three hours. How long will it take four people?

b An arrow is shot upwards from a cliff at 40 m/s. The cliff is 80 m tall. The function that describes the object's height t seconds after launch is a quadratic.

c Ayden works in an electrical shop. He earns £320 per week, plus £3 for each item he sells.

d A new company is going to make picture frames. The final area of one of the frames is 1495 cm². It will contain an A3 picture which is 30 cm by 42 cm. How wide will the outer frame be?

i $x^2 + 72x - 225 = 0$

ii $y = 3x + 320$

iii $t = \dfrac{6}{n}$

iv $s = -4.9t^2 + 40t + 80$

Graph A

Graph B

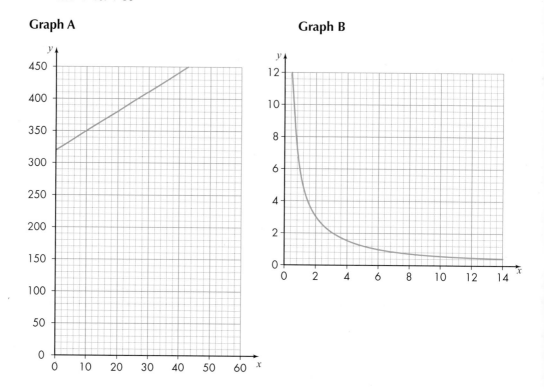

Graph C

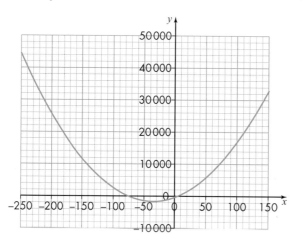

Graph D

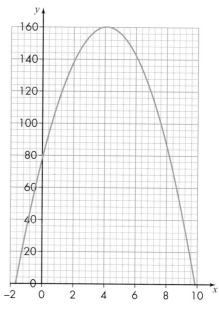

55 Look at the list of formulae.

a $y = 2x + 3$

b $z = 2(x + 3)$

c $t = -2(3 - x)$

d $z = \dfrac{-2(x + 2)}{x}$

Might it be difficult to substitute negative values into any of them? Which ones? Why? What typical mistakes might people make? Make some suggestions to help them avoid making these mistakes.

56 Explain the similarities and differences between rearranging a formula and solving an equation.

57 Without drawing the graphs, compare and contrast features of these pairs of graphs.

a $y = 2x$

   $y = 2x + 6$

b $y = x + 5$

   $y = x - 6$

c $y = 4x - 5$

   $y = -4x + 6$

d $y = 2x$

   $y = \dfrac{1}{2}x$

Use as much mathematical vocabulary as possible in your explanations.

Question	Hint
10	You may find using a few examples will help you to explain this.
22	Remember $y = mx + c$, where m is the gradient and c is the y-intercept.
27	This question can be done by many methods, but sketching a distance–time graph is the easiest. Mark a grid with a horizontal axis from 8 am to 12 pm and the vertical axis from 0 to 40 km. Draw lines for both the runner and the cyclist. Remember that the cyclist doesn't start until 9 am. What will it tell you when these lines cross?
38, 39	You may find using a few examples will help you explain this.
49	The area of the big square is $(x + 1)^2$. What about the other squares and rectangles?
54	Think about key characteristics and points on the graph.
56	You may find using a few examples will help you explain this.

Worked exemplars

1 In the table, the letters a, b, c and d represent different numbers. The total of each row is given at the side of the table.

Work out the values of a, b, c and d.

a	a	a	a	32
a	a	b	b	36
a	a	b	c	33
a	b	c	d	31

This is a problem-solving question so you need to make connections between different parts of mathematics (in this case equations and substitution) and show your strategy clearly.	
$4a = 32$ $2a + 2b = 36$ $2a + b + c = 33$ $a + b + c + d = 31$	You could start by writing out the four equations.
$4a = 32$ $\quad a = 8$	Only one of these equations can be solved immediately ($4a = 32$) because the others all include two or more variables. {| 8	8	8	8	32 |} {| 8	8	b	b	36 |} {| 8	8	b	c	33 |} {| 8	b	c	d	31 |}
$2a = 2 \times 8 = 16$ So, $16 + 2b = 36$ $\Rightarrow 2b = 20$	Now you know that $a = 8$, you can substitute it into the other equations and simplify them. It will help to work out $2a$ first. You can use the sign $\Rightarrow$ which means 'so' or 'this means that'.
$2b = 20$ $\quad b = 10$	Now you can solve $2b = 20$ and use the answer to simplify the other equations.
$16 + b + c = 33$ $b + c = 17$ b is 10, so $10 + c = 17$ $c = 7$	
$8 + b + c + d = 31$ b is 10 and c is 7, so $8 + 10 + 7 + d = 31$ and $d = 6$	
$a = 8$, $b = 10$, $c = 7$, $d = 6$	Make sure you give a full answer to the question at the end.

2 These are expressions for the nth terms of three sequences.

Sequence 1: $4n + 1$

Sequence 2: $5n - 2$

Sequence 3: $5n + 10$

Write down whether the sequences generated by the nth terms always give multiples of 5 (A), never give multiples of 5 (N) or sometimes give multiples of 5 (S).

Sequence 1 → 5, 9, 13, 17, 21, 25, 29, …	Substitute $n = 1, 2, 3, …$ until you can be sure of the sequences.
Sequence 2 → 3, 8, 13, 18, 23, 28, …	
Sequence 3 → 15, 20, 25, 30, 35, 40, …	The sign → means 'gives'.
Sequence 1: Sometimes (S)	The series generated will show whether their terms are never, sometimes or always multiples of 5.
Sequence 2: Never (N)	
Sequence 3: Always (A)	

3 Look at this number pattern.

Line 1 $\qquad\qquad 2 \times 1 = 1 \times 2$

Line 2 $\qquad\qquad 2 \times (1 + 2) = 2 \times 3$

Line 3 $\qquad\quad 2 \times (1 + 2 + 3) = 3 \times 4$

Line 4 $\quad 2 \times (1 + 2 + 3 + 4) = 4 \times 5$

a Write down line 5 of the pattern.

b Write down line 10 of the pattern.

c Use the pattern to find the sum of the whole numbers from 1 to 100.

You *must* show your working.

This question involves mathematical reasoning and problem solving. Make sure you show clearly how you reach your answers.

a $2 \times (1 + 2 + 3 + 4 + 5) = 5 \times 6$	You are extending the given patterns, so use line 4 and add '+5' into the brackets on the left, and increase the numbers in the product on the right by 1.
b $2 \times (1 + 2 + 3 + 4 + 5 + 6 + 7 + 8 + 9 + 10)$ $= 10 \times 11$	This is where you use mathematical reasoning to spot that each line number is the last value in the brackets on the left and the first value in the product on the right.
c $2 \times (1 + 2 + … + 99 + 100) = 100 \times 101$ $\therefore 1 + 2 + … + 99 + 100 = \frac{100 \times 101}{2}$ $\therefore 1 + 2 + … + 99 + 100 = 50 \times 101$ $= 5050$	This is the problem-solving part. Notice that the expression in the brackets on the left is the sum of all the numbers from 1 to n. Use the same idea as in **b** to write down the 100th line. There is no need to write out all the terms. Divide both sides by 2 and work out the answer.

4 Georgia has a card with the expression $2x + 16$.

Amy has a card with the expression $5x - 2$.

$\boxed{2x + 16}$ $\boxed{5x - 2}$

a Show that their cards are equal when $x = 6$.

b Georgia says that her card has a higher value than Amy's card when x is more than 6. Use inequalities to show that Georgia is incorrect.

This question assesses 'communicating mathematically', so whichever method you choose to answer the question, show and explain all of the steps.

a **Method 1**	There are two acceptable methods to answer part **a**. The first involves substituting $x = 6$ into both expressions and showing that they have the same value.
When $x = 6$, $$2x + 16 = 2 \times 6 + 16$$ $$= 12 + 16$$ $$= 28$$ When $x = 6$, $$5x - 2 = 5 \times 6 - 2$$ $$= 30 - 2$$ $$= 28$$ Their cards both equal 28 when $x = 6$.	
Method 2 $$2x + 16 = 5x - 2$$ $2x + 18 = 5x$ (Add 2 to both sides.) $18 = 3x$ (Subtract $2x$ from both sides.) $6 = x$ (Divide both sides by 3.)	The other acceptable method is to make the expressions equal and show that $x = 6$. Include an explanation of what you are doing.
b If Georgia is correct, then $x > 6$ when $2x + 16 > 5x - 2$. $$2x + 16 > 5x - 2$$ $2x + 18 > 5x$ (Add 2 to both sides.) $18 > 3x$ (Subtract $2x$ from both sides.) $6 > x$ (Divide both sides by 3.) This means that $x < 6$, so Georgia is incorrect.	Write the inequality from her statement and solve it to show that it is not true.

5 Mark set off for town with a £20 note and £1.50 in change.

His bus fare was 50p.

He bought three CDs that each cost the same amount and a drink that cost 98p.

He then came home on the bus, paying 50p for his fare.

Work out the most that Mark could have paid for each CD. Give your answer in pounds, correct to the nearest penny.

This is a problem-solving question, so you will need to use a series of mathematical processes to solve the problem. Remember to show your strategy clearly.

£20 + £1.50 = £21.50 $$= 2150\text{p}$$ One CD costs x p. Three CDs cost $3x$ p. Total bus fare plus drink $= 50\text{p} + 98\text{p} + 50\text{p}$ $= 198\text{p}$	Start by writing down all the information you know. Remember to convert all the amounts so that they are written in the same units. Note that Mark's money is given in pounds but what he buys is given in pence. If you use pence throughout, then there will not be any decimals until the end. Alternatively, you could convert what he buys to pounds and work with the decimal values.
$3x + 198 < 2150$	Next set up an inequality to show that three CDs plus the amount Mark spent on the other items must be less than what he set off with.

$3x < 1952$ (Subtract 198 from both sides.) $x < 650.667$ (Divide both sides by 3.)	Solve the inequality using the balancing method.
x must be less than 650.667p. The largest integer less than 650.667 is 650. The most that Mark could have paid for each CD was 650p or £6.50.	Since the question asks for the most each CD cost (and the price must be an integer), you need to work out the largest integer below the answer. Remember to give your answer in pounds.
Check by substituting £6.50 and £6.51 into your inequality. $3 \times 6.50 + 1.98 = £21.48$ $3 \times 6.51 + 1.98 = £21.51$	It is good practice to check your answer by substituting. Using £6.50, Mark spends £21.48, which is less than £21.50, but using £6.51, Mark spends £21.51, which is more than £21.50. Therefore, the CDs cannot cost more than £6.50 each.

6 Tracy and Les both drove to the airport. The distance–time graphs of their journeys are shown below.

Given that 5 miles is approximately 8 km, calculate who drove faster.

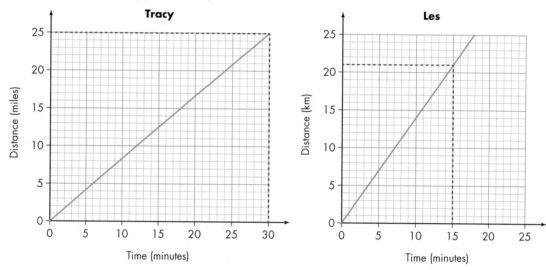

This is a problem-solving question, so you need to make connections between different parts of mathematics, in this case distance–time graphs and converting between units.	
Tracy travels 25 miles in 30 minutes, so is travelling at 50 mph. Les travels 21 km in 15 minutes, so is travelling at 84 km/h.	Start by converting the information from the graphs into speeds that can be used as a comparison. These speeds are normally given per hour.
Tracy: $50 \times \frac{8}{5} = 80$ km/h or Les: $84 \times \frac{5}{8} = 52.5$ mph	5 miles is equal to 8 km, so calculate Tracy's speed in km/h by multiplying her speed in mph by $\frac{8}{5}$. As a result, both Tracy's and Les' speeds are now in km/h. Alternatively, multiply Les' speed by $\frac{5}{8}$ to convert it to mph.
So: Tracy (80 km/h) is travelling more slowly than Les (84 km/h) or Tracy (50 mph) is travelling more slowly than Les (52.5 mph).	State your conclusion, comparing the two speeds.

3 Ratio, proportion and rates of change

1 Sunil is going to cook a leg of lamb. It needs to be cooked by 8 pm.

The cooking time for the lamb is calculated as 30 minutes per 400 g plus 20 minutes resting time.

The leg of lamb has a mass of 2 kg. Sunil plans to start cooking at 6:30 pm. Will the lamb be ready by 8 pm? Show your working.

2 Amer spends her birthday money of £60 on books and clothes in the ratio 1 : 2 . How much does she spend on clothes?

3 There are 150 cars in a carpark. 25 of the cars are green.

  a What percentage of the cars are not green?

  b One quarter of the cars are red, rounding to the nearest car. Are there more red cars than green cars?

  c Alicia says, 'Over half the cars are either red or green.' Is she right?

  d Billy says, 'Exactly $\frac{1}{3}$ of the cars are silver.' Could he be right? Explain your answer.

4 Which is longer: $\frac{1}{5}$ of a day or 360 minutes? Show your working.

5 **a** When calculating percentages of quantities, what percentage do you usually start from? How do you use this percentage to work out others? Give at least two examples.

  b To calculate 10% of a quantity, you divide it by 10. So to calculate 20%, you divide by 20. What is wrong with this statement?

6 **a** What do you look for when cancelling a ratio to its simplest form? How do you know when you have the simplest form of a ratio?

  b Give an example to show how you know when you have the simplest form of a ratio.

7 Peter, Felicity and Jack shared some money in the ratio 2 : 4 : 7. Jack got £120 more than Felicity. How much money did Peter get?

8 Are these statements correct? Explain your answers.

  a $\frac{2}{3}$ is bigger than $\frac{3}{5}$.

  b $\frac{3}{5}$ is the same as 60%.

  c 70% of £150 is more than $\frac{3}{4}$ of £150.

9 **a** In a survey of 48 students, $\frac{1}{3}$ liked football best, $\frac{1}{4}$ liked tennis best, $\frac{3}{8}$ liked athletics best and the rest liked swimming best. How many liked swimming best?

  b In year 10, half the students catch a bus to school and one-quarter cycle. 150 students either catch the bus or cycle to school. How many students are there in year 10?

10 Research has found that in an ideal coin system:

> The diameter of the coins should not be smaller than 15 mm and not be larger than 48 mm.

> The diameter of the next larger coin must be at least 25% larger than the last one.

> The machinery that makes a coins can only produce coins with diameters of a whole number of millimetres (e.g. 15 mm is allowed, 15.2 mm is not).

Design a set of coins that satisfy these requirements. Start with a 15 mm coin. Your set should contain as many coins as possible.

11 Krish is comparing bank accounts. The first offers £160 cashback when you open the account. The second offers 2% interest on balances over £2000.

Krish keeps £2500 in the account. He withdraws the interest in full at the end of each year. How many years will he need to keep the second account to make it the best choice?

12 The area of the large square (including the frame) is twice the area of the small square. The frame of the large square is $\frac{1}{7}$ of its area. The small square is divided into equal sections and three of these are shaded blue.

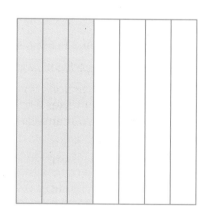

The small square is put in the middle of the large square. What fraction of the new square is shaded blue?

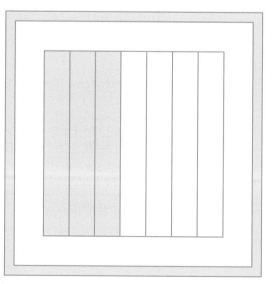

13 Anna says, 'My train fare to Sheffield increased by 15%, but then I managed to negotiate a special reduction of 15% so I'm back where I started from.'

Is Anna correct? Explain your answer.

14 Pens normally cost £1.50 each. Two shops have a special offer on.

Sian wants to buy 30 pens. Explain which shop has the best deal.

Pens-R-Us
Pay for two pens and get one free.

Budget Stationery
Pay for 5 pens and get 3 free.

15 Shampoo is on offer at two shops.

Explain which offer is the best deal.

Special offer > **500 ml 3 for 2**

Buy one Get one FREE
300 ml

16 **a** Is the ratio 1 : 6 the same as the ratio 6 : 1? Explain your answer.

b Show that 19 : 95 is the same ratio as 1 : 5.

c The instructions on a packet of cement say: mix sand and cement in the ratio 6 : 1. A builder mixes 6 kg of sand with one bucketful of cement. Could this be correct? Explain your answer.

d The ratio of boys to girls at a school club is 2 : 5. Could there be 24 students at the club altogether? Explain your answer.

17 Batteries are sold in packs.

A pack of 3 batteries costs £1.50.

A pack of 15 batteries costs £5.

A pack of 25 batteries costs £10.

a Work out the cheapest way to buy **exactly** 90 batteries.

b The packs of 15 are now available for buy two get one free. Does this change your answer?

18 The answer to a percentage increase question is £10. Make up a question with this answer.

19 **a** A golf club has 24 female members. The ratio of male to female members of a golf club is 5 : 2. How many members does the golf club have?

b Richard, Shaun and Joan are paying a restaurant bill of £85. They want to split the bill in the ratio 2 : 3 : 5. How much does Shaun pay?

c Design your own question like the one in part **a**. Try to use a different context.

20 **a** Sarah is comparing tablet prices. The battery life of the more expensive tablet is $\frac{5}{4}$ times the battery life of the cheaper tablet. The cheaper model has 8 hours of battery life. How long does the more expensive model have?

b The more expensive tablet costs £198. The cheaper one is £118. Is this less than or greater than the proportional change to the battery life? Justify your answer.

c Sarah thinks she may be able to negotiate a reduction. What reduction would she need to negotiate to make it worth buying the more expensive model based on the proportional change in battery life?

21 Jamelia wants to buy some blank CD-R discs. She is trying to decide which deal is best.

5-pack
90 minutes each
£6.50

5-pack
80 minutes each
£6.50

5-pack
80 minutes each
£4.00

a Which pack of CDs is the best buy?

b Why might someone not choose the best buy?

22 Mr and Mrs Fitzpatrick are going to Germany. They each have £400 to change into euros. They see this deal in a bank.

How much more money will they get by putting their money together before they change it?

Fantastic Rates

Get 1.19 Euros for £1 on amounts less than £500

Get 1.22 Euros for £1 on amounts greater than £500

23 Philippa makes biscuits that she sells at children's parties. She has the following ingredients.

8 kg plain white flour

3 kg caster sugar

2 kg butter

7 kg icing sugar

To make 24 birthday biscuits she needs:

250 g plain white flour

85 g caster sugar

20 g butter

2 tbsp lemon curd

250 g icing sugar

1 tbsp strawberry jam

She has plenty of jam and lemon curd.

a Philippa sells her biscuits in packs of 15. How many packs of biscuits can she make using the ingredients she has?

b Philippa sells three-quarters of her biscuits to individuals at £2.99 per pack. The rest she sells in bulk to one buyer at a discount of 15%. The ingredients cost her £59 plus an additional £26 for extras such as packing.

What is her percentage profit?

24 John is buying a new computer which is advertised for £595 exclusive of VAT. VAT is charged at 20%.

'I managed to negotiate a 20% discount so I only need to pay the shop £595.'

Do you agree or disagree with this statement? If he pays the shop £595 will he have got the deal he thinks he has? Explain your response to John.

The value of a car depreciated (decreased) by 7%. The car is now worth £11 630. What was the car worth originally?

Equate the values 93% represents £11 630

Write this as an equation using a decimal multiplier:

$A \times 0.93 = £11\,630$

Rearrange the formula to find A:

$$A = \frac{11\,630}{0.93}$$

$A = 12\,505$ to the nearest whole number.

So the original value was £12 505.

25 **a** The original value before a sale was £A. The sale reduction was 15%. Write a formula to describe this proportional change.

b Now rearrange the formula to show how you work out the original value when you know the sale price.

26 Explain your answer for each part of this question.

a Is a 50% increase followed by a 50% increase the same as doubling?

b A shop is offering a 25% discount. Is it better for the customer to have this before or after VAT is added at 20%?

27 **a** After one year the value of a scooter has depreciated by $\frac{1}{7}$ and is valued at £996. What was its value at the beginning of the year?

b Waiting staff at a restaurant get a 4% wage increase. The new hourly rate is £6.50. What was it before the increase?

c Sadie has savings of £957.65 after 7% interest has been added. What was the original amount of her savings before the interest was added?

d How do you work out a multiplier to calculate an original value after a proportional increase or decrease?

e How can you tell whether a multiplier increases or a decreases a quantity?

28 A taxi company has a fare structure based on three things: a minimum fare of £2.50, a charge for the time taken and a charge for the distance travelled.

The charge for the time taken is directly proportional to the time taken.

The charge for the distance travelled is directly proportional to the distance travelled.

Use this information about the company's fares to work out the charge per mile and the charge per minute. Then suggest a competitive pricing structure for another taxi firm.

Time taken	2 minutes	5 minutes	10 minutes	12 minutes	15 minutes
Distance	1 mile	2 miles	3 miles	5 miles	6 miles
Total charge	£2.50 (minimum fare)	£4.00	£6.50	£9.90	£12.00

29 **a** Explain why travelling a distance of 30 miles in 45 minutes gives an average speed of 40 mph.

b Explain a common mistake people might make when answering part **a** of this question.

c Explain how the units of speed help you to solve a problem. You can use your answer to part **a** to help you explain this.

30 Don wants to do some planting in his garden. He marks out a shape with an area of 2 m². However, he then decides that he wants to use more garden space. What is the area of a similar shape when the lengths are four times bigger than the corresponding lengths of the first shape?

31 A can of paint is 30 cm high and holds 5 litres of paint. How much paint does a similar can that is 75 cm high hold?

32 A sculptor has made a model statue that is 15 cm high and has a volume of 450 cm³. The real statue will be 4.5 m high. In order to buy enough materials, she needs to know the volume of the real statue. Work this out for her, giving your answer in m³.

33 Sam invests £8000 in an account for two years. The account pays 3% compound interest each year.

How much will Sam have in her account at the end of three years?

34 Each week, Raja takes out 20% of the amount in his bank account to spend. After how many weeks will the amount in his bank account have halved from the original amount?

⑦Hints and tips

Question	Hint
19b	Make sure you read the question carefully. How much would everyone pay if Richard paid £2?
19c	Think about what you did in part **a** and work backwards.

Worked exemplars

 1 Jonathan is comparing two ways to travel from his flat in London to his parents' house in Doncaster.

Tube, train and taxi

It takes 35 minutes to get to the railway station by tube in London.

The train journey from London to Doncaster takes 1 hour 40 minutes.

From Doncaster, it is 15 miles by taxi at an average speed of 20 mph.

Car

The car journey is 160 miles at an average speed of 50 mph.

Which journey takes longer: tube, train and taxi or car?

This question assesses 'communicating mathematically', so you must display your methods clearly and include words to explain what your calculations show. Imagine that you will pass your answer to a friend or relative and ask yourself whether or not they could understand it.	
For the taxi time = distance ÷ speed $= \dfrac{15}{20}$ = 0.75 hour (or 45 minutes)	First, work out the time taken by the taxi. It is not essential to show the formula 'time = distance ÷ speed' but it is useful to draw the triangle that shows the relationship. Be careful with time as a decimal: 0.75 hours = 45 minutes.
Total time for the journey by tube, rail and taxi: 35 minutes + 1 hour 40 minutes + 45 minutes = 3 hours	Next, work out the total time for tube, train and taxi.
For the car time = distance ÷ speed $= 160 ÷ 50$ = 3.2 hours (or 3 hours 12 minutes)	Now work out the time taken by the car. 0.1 hours = 6 minutes so 0.2 hours = 12 minutes
Travelling by car takes 12 minutes longer.	Finally, compare the times taken. It is essential to write a final conclusion: do not assume it is obvious from your working.

2 Which of the following ratios in not the same as 3 : 5?

$$1 : 1\tfrac{2}{3} \qquad 1 : \tfrac{8}{5} \qquad 9 : 15 \qquad 39 : 65$$

This is a reasoning question asking you to work out which ratio is not the same as 3 : 5.	
Eliminate to leave $1 : 1\tfrac{2}{3}$ or $1 : \tfrac{8}{5}$	Eliminate 9 : 15 and 39 : 65 as these will cancel down to 3 : 5.
The ratio 3 : 5 is the same as $1 : 1\tfrac{2}{3}$.	To get 1 as the first value divide by 3, 5 ÷ 3 = 1.666.
So $1 : \tfrac{8}{5}$ is not the same as 3 : 5.	Check by multiplying both sides of the ratio by 3 to get 3 : 4.6, which is not 3 : 5.

3 A, B, C and D are four points on a number line.

$AB : BC = 7 : 3$

$BC : CD = 2 : 5$

Work out the ratio $AC : CD$.

Give your answer in its simplest form.

This question requires you to 'interpret and communicate information', so you will need to extend the information beyond what is stated explicitly.	
	It may help to draw a diagram.
This table shows the given ratios. 	AB	7	
BC	3	2
CD		5		Set up a table and write in the information you are given.
BC has the same value in the first and second columns. 	AB	7		14
BC	3	2	6
CD		5	15		Now extend and complete the third column. Multiply AB and BC in the first column by 2 to give AB = 14 and BC = 6. Multiply BC and CD in the second column by 3 to give BC = 6 and CD = 15.
Add AB and BC to find AC. 	AB	7		14	20	4
BC	3	2	6		
CD		5	15	15	3		Now extend and complete a fourth column by combining AB and BC from the third column to get AC = 20 (column 4) and cancel AC and CD by a factor of 5 (column 5).
The ratio of AC : CD is 4 : 3.	Give your answer.

4 $A : B = 4 : 5$. What fraction of B is A?

$\dfrac{4}{5}$ $\qquad$ $\dfrac{4}{9}$ $\qquad$ $\dfrac{5}{9}$ $\qquad$ $\dfrac{5}{4}$

This is a problem-solving question. The way the question is presented is designed to distract you.	
Let A be 4 and B be 5.	Choose the simplest possible values for A and B.
$\dfrac{4}{5}$	Check by dividing both sides by 5 to give B as 1. So the ratio is $\frac{4}{5} : 1$, which shows that A is $\frac{4}{5}$ of B.

5 Laura buys a house for £240 000.

She pays a deposit of 20%.

She takes out a 25-year mortgage for the rest of the cost.

She will have to pay 0.525% of the mortgage each month.

How much more than £240 000 will the house eventually cost?

This is a problem-solving question where you need to translate a real-life problem into a series of mathematical steps. Show your working clearly and use words to explain what you are calculating.	

Deposit   = 0.2 × £240 000
              = £48 000
 	
Mortgage = £240 000 – £48 000
              = £192 000
	Work out how much the mortgage will be.

Monthly cost = £192 000 × 0.525 ÷ 100 = £1008	Work out the monthly cost of the mortgage.

Total mortgage repayments
              = 25 × 12 × £1008
              = £302 400
	Work out the total of the mortgage repayments.

Total amount paid = £302 400 + £48 000 = £350 400	Work out the total amount she pays.
Extra money paid = £350 400 – £240 000 = £110 400	Subtract the cost of the house from the total amount she paid. You could reduce the steps in your calculation by subtracting the initial loan amount from the total mortgage repayments: 302 400 – 192 000 = £110 400.

6 Two cubes have the same mass.

Cube A has a side of 2 cm and is made from a material with a density of 34 g/cm^3.

Cube B has a side of 3 cm.

Work out the density of the material in cube B.

Give your answer to the nearest gram per cubic centimetre.

In this question you need to interpret and communicate information accurately.	
Mass of cube A: 2^3 × 34 = 272 g	Work out the mass of cube A. Remember, the volume of a cube is the cube of the length of the side, which is 2^3 in this case.
Density of cube B: 272 ÷ 3^3 = 10.074 g/cm^3	Work out the density of Cube B. Divide the mass by the volume.
Density of cube B is 10 g/cm^3.	Round to the nearest g/cm^3.

7 Kelly saved just enough money to buy the new TV that she wanted. When she got to the shop to buy the TV, there was a sale on and the price was reduced by 15% to £319.60. She sees a radio that costs £42.50. Will she have enough money left to buy the radio after she has paid for the TV?

This is a communicating mathematically question, so you must show your working clearly and explain in words what you are working out.	
The price of the TV is reduced by 15%, so the multiplier is 0.85. Original cost of TV is $319.60 \div 0.85 = 376$.	Work out the original cost of the TV to find the amount of money that Kelly has saved.
Kelly will have $376 - 319.60 = £56.40$ after buying the TV.	Work out how much she will have left after buying the TV.
$56.40 > 42.50$, so she will have enough to buy the radio.	Write a clear conclusion.

8 A plant in a greenhouse is 10 cm high. Its height increases by 13% each day. How many days does it take to double in height?

This is a problem-solving question. You need to find a way to work out the number of days the plant will take to double its height. This can be solved using more than one method. The trial and improvement method is shown here.	
The multiplier for a 13% increase is 1.13.	Write down the multiplier.
Try 5 days. $10 \times 1.13^5 = 18.42$	Try a number of days. The result is too small, so try a bigger number next.
Try 8 days. $10 \times 1.13^8 = 26.58$	This is too big, so try a number between your first guess and your second guess.
Try 6 days. $10 \times 1.13^6 = 20.82$	This is approximately correct.
The plant doubles in size in 6 days.	Write a conclusion.

9 Which of these calculations would give the answer to the value, after 2 years, of an investment of £6000, which increased by 5% in the first year and then 4% in the second year?

6000×1.09 $6000 \times 1.05 \times 1.04$ 6000×2.09 $6000 \times 1.05 + 6000 \times 1.04$

This is a mathematical reasoning question with four choices. You can either eliminate those that are clearly wrong or use your knowledge of percentage change to work out the answer. All the four answers use multipliers.	
6000×1.05	Write down the calculation that gives the value after one year. There is no need to work it out.
$(6000 \times 1.05) \times 1.04$	This value is increased by 4% (a multiplier of 1.04) in the second year.
$6000 \times 1.05 \times 1.04$	Simplify the answer to give the second option.

4 Geometry and measures

1   A point P has coordinates (a, b).

　　a　The point P is rotated 90° clockwise about (0, 0) to give a point Q. What are the coordinates of Q?

　　b　The point P is rotated 180° clockwise about (0, 0) to give a point R. What are the coordinates of R?

　　c　The point P is rotated 90° anticlockwise about (0, 0) to give a point S. What are the coordinates of S?

2   A designer is making a logo for a company. She starts with a kite ABCD. She then reflects the kite in the line BD on top of the original kite to create the logo.

　　Draw a kite on squared paper. Use the designer's method to draw the logo.

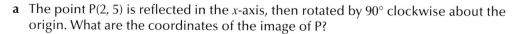

3   a　The point P(2, 5) is reflected in the x-axis, then rotated by 90° clockwise about the origin. What are the coordinates of the image of P?

　　b　The point Q(a, b) is reflected in the x-axis, then rotated by 90° clockwise about the origin. What are the coordinates of the image of Q?

4   a　The point R(4, 3) is reflected in the line $y = -x$, then reflected in the x-axis. What are the coordinates of the image of R?

　　b　The point S(a, b) is reflected in the line $y = -x$, then reflected in the x-axis. What are the coordinates of the image of S?

5   Triangle A has coordinates (2, 2), (6, 2) and (6, 4).

　　Triangle A is enlarged by a scale factor of $\frac{1}{2}$ about the origin to give triangle B.

　　What are the coordinates of triangle B?

6   a　Describe a rectangle precisely so someone else can draw it. What mathematical words are important?

　　b　Does a square meet the definition of a rectangle? Explain why or why not.

　　c　Describe what properties you need to be sure a triangle is:

　　　　i　isosceles

　　　　ii　equilateral

　　　　iii　scalene.

7 **a** What properties do you need to know about a quadrilateral to be sure it is:

    i a kite

    ii a parallelogram

    iii a rhombus

    iv an isosceles trapezium?

  b Show that a rhombus must be a parallelogram but a parallelogram is not necessarily a rhombus.

  c Why can a trapezium not have three acute angles?

  d Which quadrilateral can have three acute angles?

8 **a** Sketch a quadrilateral that has:

    i one line of symmetry

    ii two lines of symmetry

    iii three lines of symmetry

    iv no lines of symmetry.

  b What is the order of rotational symmetry of each of the quadrilaterals you sketched?

9 The diagram shows a compound shape made up of rectangles.

  a Explain how you work out the perimeter.

  b Show two different ways of working out the area of the shape.

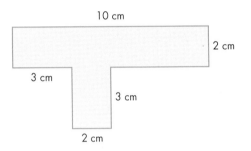

10 **a** The length of the diagonal of a square is 20 cm.

    What is the perimeter of the square?

  b Joe is told that the diagonal of a square is 8 cm.

    He says, 'In that case, the area of the square must be 32 cm².'

    Explain how he worked this out.

11 State what changes and what stays the same about a shape when you:

  a translate it

  b rotate it

  c reflect it.

12 Give two reasons why the trapezium is different from the parallelogram.

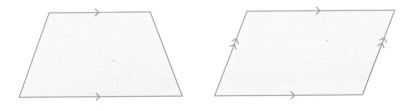

13 An isosceles triangle has one angle of 30°. Is this enough information to know the sizes of the other two angles? Why?

14 Use what you know about triangles and any information in these diagrams to justify why each of the triangles could be the odd one out.

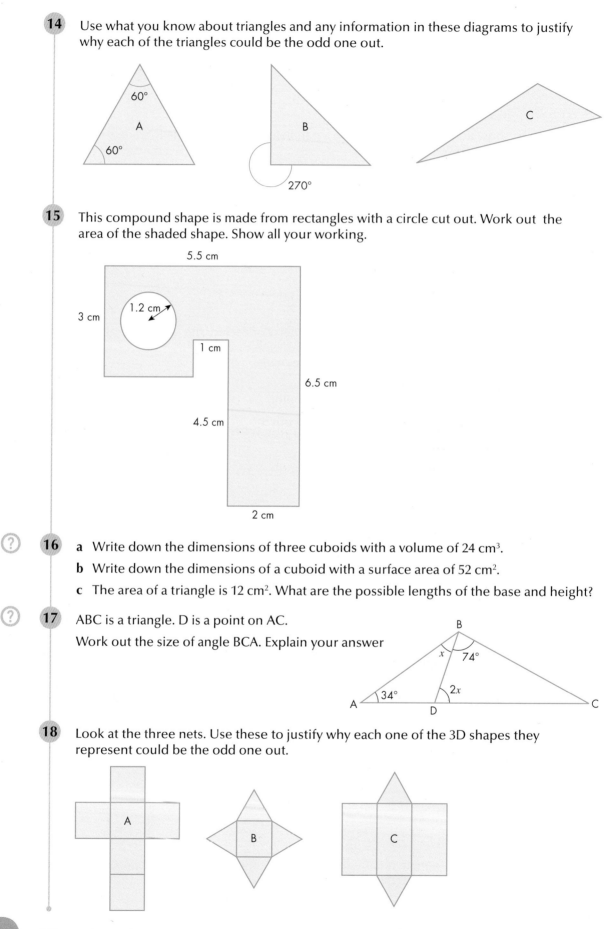

15 This compound shape is made from rectangles with a circle cut out. Work out the area of the shaded shape. Show all your working.

16 a Write down the dimensions of three cuboids with a volume of 24 cm³.

   b Write down the dimensions of a cuboid with a surface area of 52 cm².

   c The area of a triangle is 12 cm². What are the possible lengths of the base and height?

17 ABC is a triangle. D is a point on AC.

   Work out the size of angle BCA. Explain your answer

18 Look at the three nets. Use these to justify why each one of the 3D shapes they represent could be the odd one out.

19 The diagram shows a compound shape. Write down three different ways of working out its area.

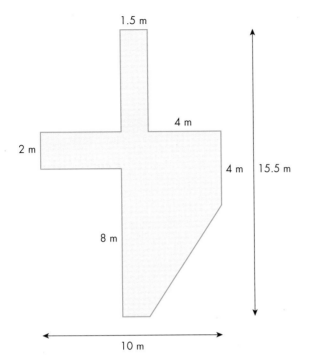

20 Is each statement true or false? Explain why. If the statement is false, explain what mistake has been made.

a The perimeter of this shape is 18 cm.

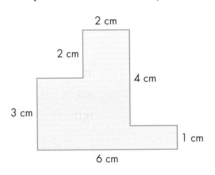

b $\frac{1}{5}$ of this shape is shaded.

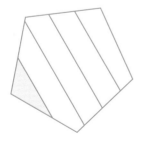

21 Tristan needs to put tables in rows in a conference room. The room is 20 m by 18 m. The tables are 2 m by 3 m. There needs to be at least 1.5 m around each table for seating and movement. The stage is 10 m by 4 m.

Each table seats 7 people in a way that they can see the stage. Eighty people are coming to the conference. Will he be able to seat them all?

Show your working to justify your answer.

22 What do you need to know to be able to work out the volume and surface area of a cylinder?

23 A tower, CD, is at the top of a hill, BC. Kamal is a surveyor and needs to work out the height of the tower. He measures the distance AC as 70 m and the angles of elevation of the top and bottom of the tower as 25° and 42°, respectively.

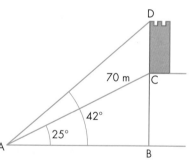

Calculate:

a the angle CAD

b the length AB

c the length CB

d the height of the tower, CD.

24 The diagram shows a plan of two circular lawns. Which is the correct area of the lawn? Explain your answer as well as what is wrong for the other answers.

80π m²

208π m²

24π m²

48π m²

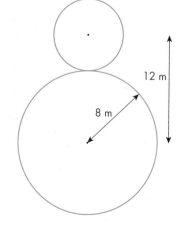

25 The diagram shows a rectangle ABCD.

Use a straight edge and a pair of compasses to construct a triangle twice the area of the rectangle ABCD.

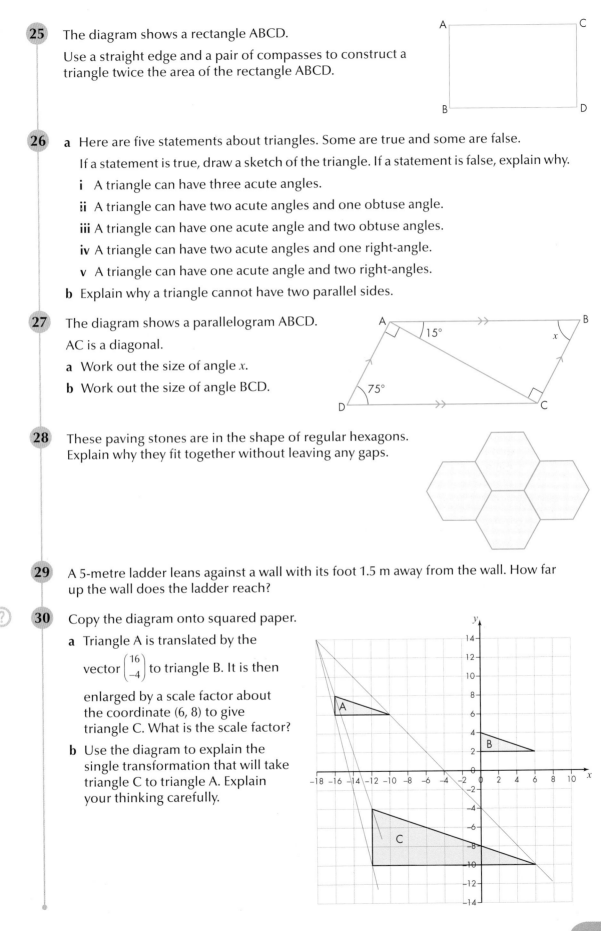

26 **a** Here are five statements about triangles. Some are true and some are false.

If a statement is true, draw a sketch of the triangle. If a statement is false, explain why.

   i A triangle can have three acute angles.

   ii A triangle can have two acute angles and one obtuse angle.

   iii A triangle can have one acute angle and two obtuse angles.

   iv A triangle can have two acute angles and one right-angle.

   v A triangle can have one acute angle and two right-angles.

  b Explain why a triangle cannot have two parallel sides.

27 The diagram shows a parallelogram ABCD.

AC is a diagonal.

  a Work out the size of angle x.

  b Work out the size of angle BCD.

28 These paving stones are in the shape of regular hexagons.
Explain why they fit together without leaving any gaps.

29 A 5-metre ladder leans against a wall with its foot 1.5 m away from the wall. How far up the wall does the ladder reach?

30 Copy the diagram onto squared paper.

  a Triangle A is translated by the

    vector $\begin{pmatrix} 16 \\ -4 \end{pmatrix}$ to triangle B. It is then

    enlarged by a scale factor about the coordinate (6, 8) to give triangle C. What is the scale factor?

  b Use the diagram to explain the single transformation that will take triangle C to triangle A. Explain your thinking carefully.

31 When you draw a semicircle on each side of a right-angled triangle, as shown in the diagram, what can you say about the areas of the circles?

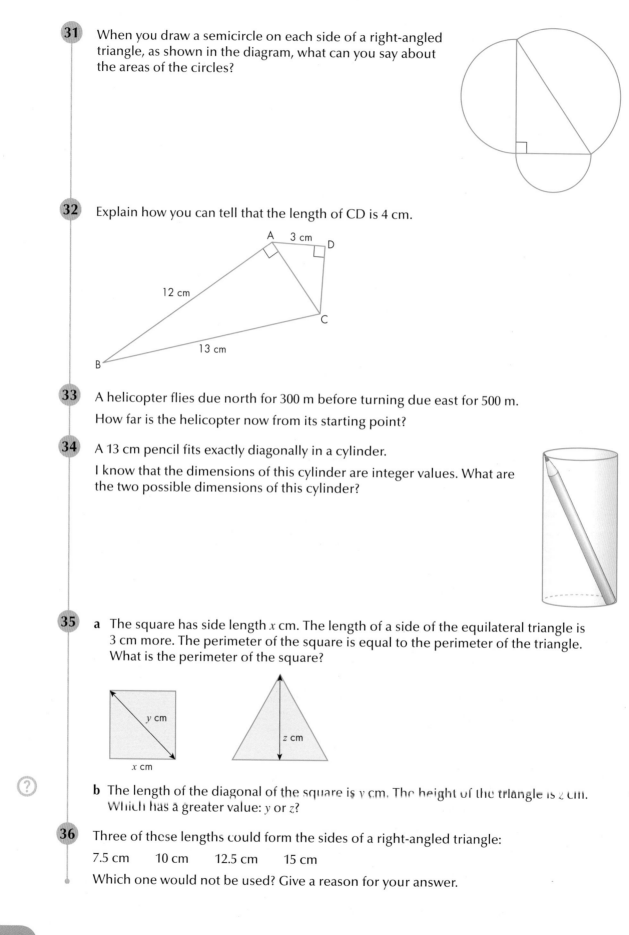

32 Explain how you can tell that the length of CD is 4 cm.

33 A helicopter flies due north for 300 m before turning due east for 500 m.

How far is the helicopter now from its starting point?

34 A 13 cm pencil fits exactly diagonally in a cylinder.

I know that the dimensions of this cylinder are integer values. What are the two possible dimensions of this cylinder?

35 **a** The square has side length x cm. The length of a side of the equilateral triangle is 3 cm more. The perimeter of the square is equal to the perimeter of the triangle. What is the perimeter of the square?

b The length of the diagonal of the square is y cm. The height of the triangle is z cm. Which has a greater value: y or z?

36 Three of these lengths could form the sides of a right-angled triangle:

7.5 cm   10 cm   12.5 cm   15 cm

Which one would not be used? Give a reason for your answer.

37 Philippa and Carl are looking at a stained glass window. Philippa says, 'It's amazing how they can create such a window with congruent shapes.'

Carl replies, 'Yes, you are right.'

How can this statement be correct?

38 The area of a rectangle is 60 cm². How many rectangles, with an area of 60 cm², with whole-number dimensions can you list?

39 The diagram shows a shaded quadrilateral inside a square.

Work out the area of the shaded quadrilateral.

Show your working.

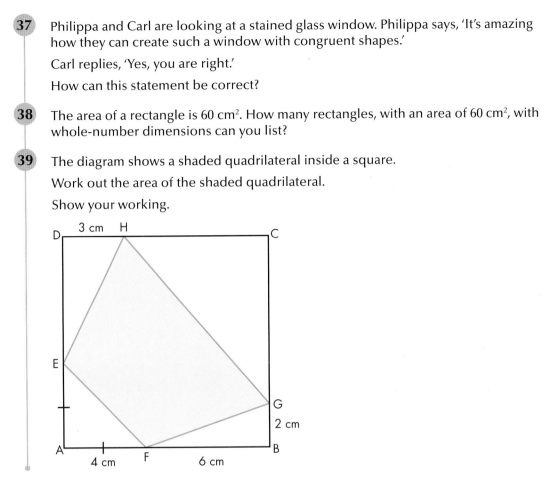

Real-life problems are often quite complicated and use a lot of words. It is important you are able to understand the words so you can find the mathematics. The following question is an example of this. Remember to think carefully what the words mean and what mathematics you will need to use. It is also often useful to highlight important pieces of information.

40 The diagram shows the plan of Mrs Williams' garden. The size of the garden is a rectangle 6.5 m wide × 4.8 m wide.

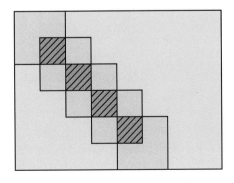

The blue area is made up of squares with sides of 1.6 m. These are going to be flower beds and need to be filled with soil.

The hatched area is made up of square paving stones 0.8 m wide and this part of the flower bed will not need soil. The green area of the garden will be grass.

Mrs Williams needs to buy enough topsoil to fill the flower beds and enough grass seed to cover the lawn.

Soil is sold in bags containing 1000 kg of soil. The cost per bag is £73.30.

The volume of 1000 kg of soil is about 0.75 m³.

Mrs Williams wants the soil to be approximately 0.5 m deep.

Grass seed is sold in 500 g bags which cost £14.99. You need approximately 50 g of grass seed per square metre.

How much will it cost Mrs Williams to complete this part of the garden?

41 Is the following statement always, sometimes or never true? Justify your answer.

When one rectangle has a larger perimeter than another one, it will also have a larger area.

42 Right-angled triangles have half the area of the rectangle with the same base and height. Is this also true for non-right-angled triangles? Justify your answer.

43 Triangle A is drawn on a grid. Triangle A is rotated to form a new triangle B.

The coordinates the vertices of B are (3, –1), (1, –4) and (3, –4).

Describe fully the rotation that maps triangle A onto triangle B.

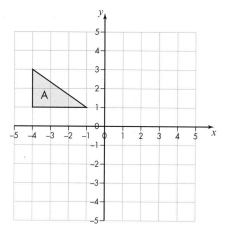

44 A council wants the outside of a block of flats repainted. The windows and doors do not need painting. A builder has been asked to provide a quote for painting the building.

The front and back of the building are 12 m high by 25 m wide. The sides of the building are 12 m high by 12 m wide.

There are 40 openings for windows and doors. Each measures 2 m by 1 m.

There will be two coats of paint. Cans of paint cost £25 and contain 10 litres of paint.

1 litre of paint will cover 16 m².

The builder thinks it will take two weeks to do and he will need three painters.

Each painter costs £120 per day.

The scaffolding costs £500.

The builder adds 10% to all the costs to cover his time.

What should the builder charge the council, including 20% VAT?

45 This is a solid cube side length 4 m. A cylinder with diameter 2.4 m is cut through from the front face to the back face.

a What is the remaining surface area of the faces of the cube?

b What is the volume of the shape?

c The outside of the cube is going to be painted light blue. The inside of the cylinder cut through the cube will be painted dark blue. 1 litre of paint covers 9 m².

How much of each colour paint will be needed?

46 Shehab says, 'As long as I know two sides of a triangle and the angle between them then I can draw the triangle.'

Is Shehab correct? If not, explain why not.

47 You are asked to construct a triangle with sides 9 cm, 10 cm and an angle of 60°. Sketch all the possible triangles that you could construct from this description.

48 The locus of a point is described as 5 cm away from point A and equidistant from points A and B. Which of the following could be true? Explain your answer.

a The locus is an arc.

b The locus is just two points.

c The locus is a straight line.

d The locus is none of these.

49 How does knowing the sum of the interior angles of a triangle help you to work out the sum of the interior angles of a quadrilateral? Will this work for all quadrilaterals?

50 One of the lines of symmetry of a regular polygon goes through two vertices of the polygon. Explain why the polygon must have an even number of sides.

51 Sketch a shape to show that:

a a trapezium might not be a parallelogram

b a trapezium might not have a line of symmetry

c every parallelogram is also a trapezium.

52 Is this statement sometimes, always or never true? Explain your answer.

The sum of the exterior angles of a polygon is 360°.

53 The angles in a triangle are in the ratio 6 : 5 : 7.

Work out the sizes of the three angles.

54 Explain why equilateral triangles, squares and regular hexagons will tessellate on their own, but other regular polygons will not.

55 What is the minimum information you need about a triangle to be able to calculate all three sides and all three angles?

56 Are these statements true or false? Justify your answers.

a Every rhombus is a parallelogram, and a rhombus with right angles is a square.

b A rhombus must be a parallelogram but a parallelogram is not necessarily a rhombus.

c A trapezium cannot have three acute angles.

d A quadrilateral can have three acute angles.

57 How do you decide whether you need to use a trigonometric relationship (sine, cosine or tangent) or Pythagoras' theorem to solve a triangle problem?

58 **a** **i** Make up a simple reflection.

    ii Make up a more complicated reflection. What makes it harder than your first one?

b **i** Make up a simple rotation.

    ii Make up a more complicated rotation. What makes it harder than your first one?

59 **a** What changes when you enlarge a shape? What stays the same? Give an example.

   b What information do you need to complete a given enlargement?

   c How do you work out the centre and the scale factor of an enlargement?

60 **a** How can you tell if a shape has been reflected or translated?

   b Describe how the image produced by rotating a rectangle about its centre looks different from the image produced by rotating it about one of its vertices.

61 The cross section of a skirting board is the shape of a rectangle, with a quadrant (quarter circle) on the top. The skirting board is 1.5 cm thick and 6.5 cm high. Lengths of board totaling 120 m are ordered. What volume of wood is in the order?

62 Starting from this 2D representation of a 3D shape:

   a How many faces will the 3D shape have? How do you know?

   b Which face will be opposite face number 2 in the 3D shape? How do you know?

   c How would you draw the plan and elevation for the 3D shape made from this net?

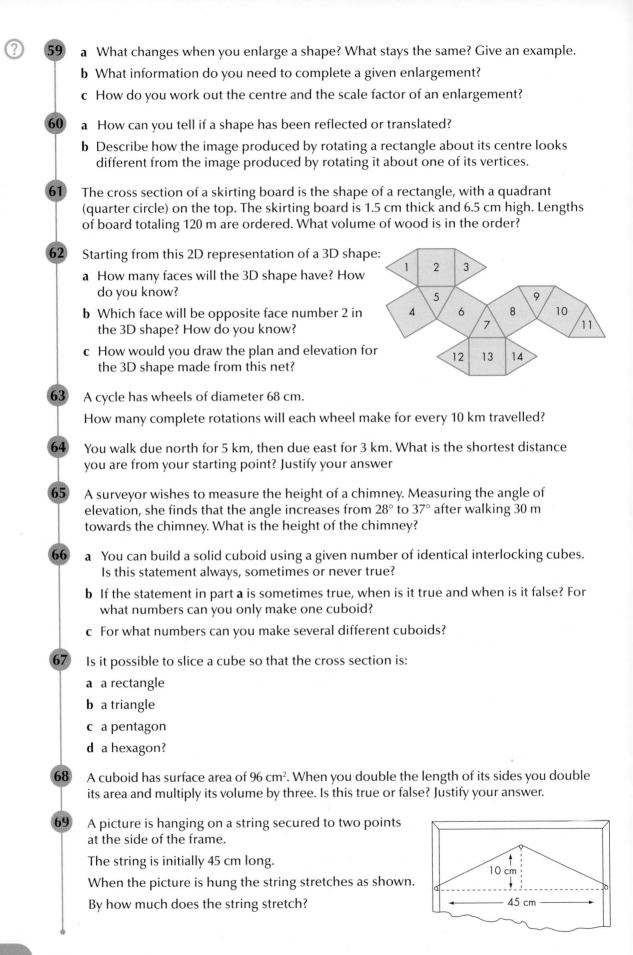

63 A cycle has wheels of diameter 68 cm.

   How many complete rotations will each wheel make for every 10 km travelled?

64 You walk due north for 5 km, then due east for 3 km. What is the shortest distance you are from your starting point? Justify your answer

65 A surveyor wishes to measure the height of a chimney. Measuring the angle of elevation, she finds that the angle increases from 28° to 37° after walking 30 m towards the chimney. What is the height of the chimney?

66 **a** You can build a solid cuboid using a given number of identical interlocking cubes. Is this statement always, sometimes or never true?

   b If the statement in part **a** is sometimes true, when is it true and when is it false? For what numbers can you only make one cuboid?

   c For what numbers can you make several different cuboids?

67 Is it possible to slice a cube so that the cross section is:

   a a rectangle

   b a triangle

   c a pentagon

   d a hexagon?

68 A cuboid has surface area of 96 cm². When you double the length of its sides you double its area and multiply its volume by three. Is this true or false? Justify your answer.

69 A picture is hanging on a string secured to two points at the side of the frame.

   The string is initially 45 cm long.

   When the picture is hung the string stretches as shown.

   By how much does the string stretch?

70 Three towns, A, B and C, are joined by two roads, as in the diagram. The council wants to build a road that runs directly from A to C. How much distance will the new road save? Give your answer to one decimal place.

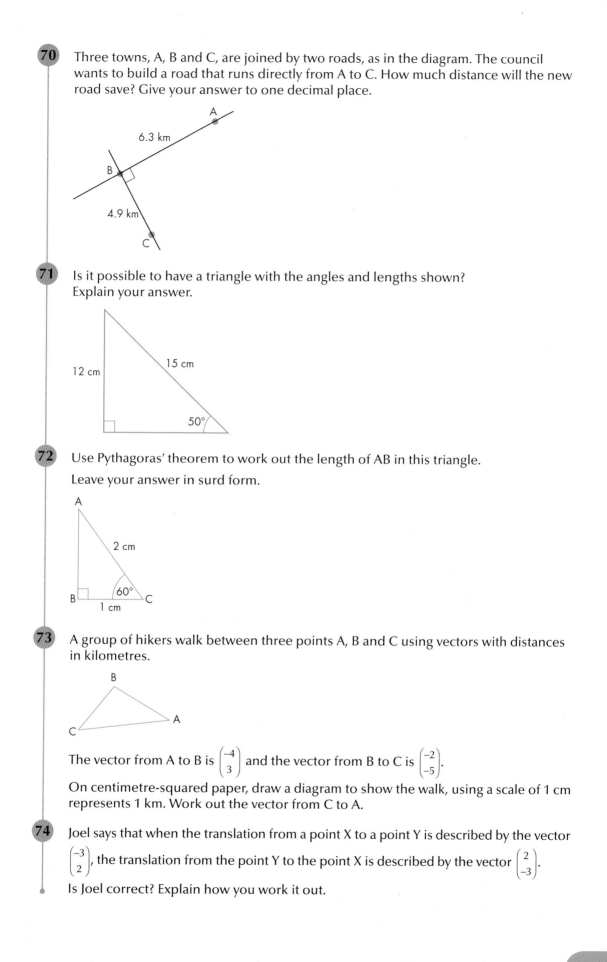

71 Is it possible to have a triangle with the angles and lengths shown? Explain your answer.

72 Use Pythagoras' theorem to work out the length of AB in this triangle. Leave your answer in surd form.

73 A group of hikers walk between three points A, B and C using vectors with distances in kilometres.

The vector from A to B is $\begin{pmatrix} -4 \\ 3 \end{pmatrix}$ and the vector from B to C is $\begin{pmatrix} -2 \\ -5 \end{pmatrix}$.

On centimetre-squared paper, draw a diagram to show the walk, using a scale of 1 cm represents 1 km. Work out the vector from C to A.

74 Joel says that when the translation from a point X to a point Y is described by the vector $\begin{pmatrix} -3 \\ 2 \end{pmatrix}$, the translation from the point Y to the point X is described by the vector $\begin{pmatrix} 2 \\ -3 \end{pmatrix}$.

Is Joel correct? Explain how you work it out.

⑦Hints and tips

Question	Hint
1	Draw a coordinate grid. Draw a point. Rotate the point 90° clockwise. What do you notice about the coordinates of the two points?
16	Think about factors.
17	Start by writing everything you know about the angles on the diagram. Don't worry to start with whether you will use it. Then decide which pieces of information will help you. Make sure you answer the question being asked!
19	Try to use as many different shapes as possible.
21	Read the question carefully and decide where the tables can go. A diagram might help.
30	An enlargement using a negative scale factor is similar to an enlargement using a positive scale factor, but the image is on the other side of the centre of enlargement, and it is upside down.
35b	You will need to use Pythagoras' theorem and make sure you check whether or not you are looking for the hypotenuse or if you need to rearrange the equation.
52	How can you use the fact that the sum of the angles on a straight line is 180° to explain why the angles at a point add up to 360°?
59	A diagram might help your justification.

Worked exemplars

1 ABC is a triangle. D is a point on AB such that BC = BD.

a Work out the value of *x*.

b Work out the value of *y*.

c Is it true that AD = DC? Give a reason for your answer.

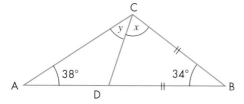

In this question you are required to communicate mathematically. You need to show clearly how you have found the missing angles and explain your final response to part **c**.

a Triangle BCD is isosceles, so angle BDC is equal to *x*. Angles in a triangle = 180° Therefore, $x + x + 34° = 180°$ $\qquad\qquad\quad 2x = 146°$ $\qquad\qquad\quad\ x = 73°$	First, make an equation in *x* from your knowledge that angles in a triangle add up to 180°. Then solve the equation.
b **Method 1** Angle ADC = 180° − 73° $\qquad\qquad = 107°$ (*angles on a line*) $y + 38° + 107° = 180°$ (*angles in a triangle*) $\qquad y + 145° = 180°$ $\qquad\qquad\ y = 35°$ **Method 2** Angle ACB = 180° − (38° + 34°) $\qquad\qquad\ = 108°$ (*angles in a triangle*) $y + x =$ Angle ACB $\qquad\ = 108°$ $\ y = 108° − 73°$ $\ y = 35°$	To work out angle *y*, you need to show how you are using the given angles and the found angle *x*. You should show the mathematical reasoning used at each stage. There are two ways of working out *y* here. Both are acceptable.
c No, *y* is not 38° so triangle ACD is not an isosceles triangle. No two sides of the triangle are equal.	Clearly state your explanation about ACD not being isosceles. The answer 'No' alone, is not enough.

2 A shape is made up of two squares and an equilateral triangle as shown. Show that the area of the shape is between 50 cm² and 75 cm².

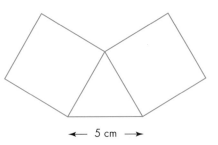

← 5 cm →

This question assesses 'communicating mathematically', so you must display your methods clearly and include words to explain what your calculations show.	
Area of one square = 5 × 5 = 25 cm² Area of two squares = 50 cm² So, the area of the shape must be greater than 50 cm².	You need to recognise that you can calculate the accurate area of the squares but not the area of the triangle.
The area of the triangle must be less than the area of one of the squares.	You do not need to calculate the area of the triangle, only show that it must be less than the area of a square.
So, the area of the shape must be less than the area of three squares, which is 75 cm².	Write a conclusion for your working.

3 Work out the single transformation that is equivalent to a rotation of 90° clockwise about the origin, followed by a reflection in the line $y = x$.

This is a problem-solving question so you will need to show your strategy.	
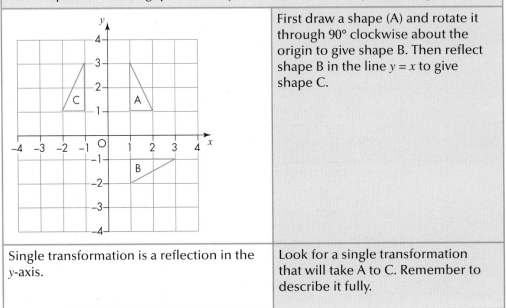	First draw a shape (A) and rotate it through 90° clockwise about the origin to give shape B. Then reflect shape B in the line $y = x$ to give shape C.
Single transformation is a reflection in the y-axis.	Look for a single transformation that will take A to C. Remember to describe it fully.

4 A baker uses square and circular tins to make his cakes.

8.5 cm

17.5 cm

8.5 cm

19 cm

He thinks that the square tin will hold more cake mix. Show that he is correct.

This question assesses 'communicating mathematically', so you must display your methods clearly and include words to explain what your calculations show and why this means the baker is correct.	
Volume of the square tin $= lwh$ $= 17.5 \times 17.5 \times 8.5$ $= 2603.125 \text{ cm}^3$ Diameter of circular tin is 19 cm, so radius $= 9.5$ cm. Volume of the circular tin $= \pi r^2 h$ $= \pi \times 9.5^2 \times 8.5$ $= 2409.9943 \text{ cm}^3$	You need to show clearly how you calculate the volume of each shape. There is no need to round the answers it is only important to know which one is larger.
$2603.125 > 2409.9943$, so the baker is correct that the square tin will hold more cake mix than the circular tin.	Finish by stating that the baker is correct and how your working shows this.

5 The key hole shape shown is made up of a circle of radius 1 cm and a sector of angle 30°.

Show that the area of this shape is 7.1 cm² (2 sf).

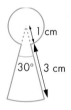

1 cm

30° 3 cm

This is a communicating mathematics question where you need to construct and communicate chains of reasoning to achieve a given result.	
Area of sector $= \dfrac{\theta}{360} \times \pi \times r^2$ $= \dfrac{30}{360} \times \pi \times 4^2$ $= \dfrac{480}{360} \times \pi$ $= 4.188\ 790\ 205$	Plan your solution by considering how to work out the total area of the shape. Remember that in this type of question you have the final result to work towards, so if your answer is not correct, look again to see what you have done wrong.
Area of major sector in circle $= \dfrac{\theta}{360} \times \pi \times r^2$ $= \dfrac{330}{360} \times \pi \times 1^2$ $= 2.879\ 793\ 266$	Write down all the digits in your calculator display. You need to show that when you add the two areas together, the final answer rounds to the given answer to 2 sf.
Total area $= 4.188\ 790\ 205 + 2.879\ 793\ 266$ $= 7.068\ 583\ 471$ $= 7.1$ (2 sf)	

6 **a** Calculate the area of a regular hexagon of side 6 cm.

   b Comment on the accuracy of your answer.

This is an evaluating question where you are required to comment on a result.	
a Base of right-angled triangle is $\frac{1}{2}$ of 6 cm = 3 cm Using Pythagoras' theorem: (height of triangle)^2 + 3^2 = 6^2 (height of triangle)^2 = 6^2 − 3^2 height of triangle = $\sqrt{(6^2 - 3^2)}$ Area of one triangle = $\frac{1}{2}$ × 6 × height Area of hexagon = 6 × area of triangle $\quad = 6 \times \frac{1}{2} \times 6 \times \sqrt{(6^2 - 3^2)}$ $\quad = 93.530743...$ $\quad = 94$ cm^2 (2 sf)	You need to show how you have divided the hexagon into six equilateral triangles and then divided one of these triangles in half to find a right-angled triangle. You need to show how you are accurately calculating the area of the shape without rounding too early. You could simplify to $\sqrt{27}$, but this isn't necessary. Then give a final answer with suitable rounding.
b The accuracy was kept by not rounding until the last stage. The initial data was assumed to be accurate, and so 2 sf gives an appropriate degree of accuracy.	You should make a suitable comment reflecting the accuracy, giving a clear reason why you selected the accuracy you did.

7 AB and CD are parallel.

E is the midpoint of AD.

Show that triangle ABE is congruent to triangle CDE.

This is a communicating mathematics question so you must show clear reasons at each stage.	
	Identify the elements that are identical in both triangles. It can help to draw out the two triangles separately and match up the sides and angles this way.
AE = DE (E is midpoint of AD) ∠BAE = ∠CDE (alternate angles) ∠AEB = ∠CED (opposite angles)	Clearly state the reason why each pair of sides or angles is identical.
So △ABE ≡ △CDE (ASA).	Finish with the clear statement that you have used ASA to show congruency.

5 Probability

1 Give examples of probabilities (as percentages) for events that could be described using each phrase:

   a impossible

   b almost (but not quite) certain

   c fairly likely

   d an even chance.

2 Make up an example of a situation with equally likely outcomes with probabilities of $\frac{1}{3}$. Justify why your example works.

3 What words would you use to describe an event with a probability of:

   a 80%

   b 0.3?

4 When you spin a coin, the probability of getting a head is 0.5. Dylan says that this means if you spin a coin 10 times you will always get exactly five heads. Is he correct? Explain your answer.

5 Make up an example of a situation with equally likely outcomes with each of these probabilities: 0.5, $\frac{1}{5}$, 0.6, 25%. Justify your answers.

6 Dean says there are three possible outcomes for a football match: win, lose or draw. Therefore the probability of winning must be one-third. Explain why he is wrong.

7 Give an example of two mutually exclusive events.

8 Is each statement true or false? Justify your answers.

   a Experimental probability is more reliable than theoretical probability.

   b Experimental probability gets closer to the true probability as more trials are carried out.

   c Relative frequency finds the true probability.

9 Sarah spins a coin 100 times and counts the number of times she gets a head. A computer is programmed to spin a coin 10 000 times. Which is most likely to be closer to getting an equal number of heads and tails? Explain why.

10 Give two events that could be described as independent. Justify your answer.

11 Give an example of what is meant by equally likely outcomes. Explain your answer.

12 Give an example of an event for which the probability can only be calculated through an experiment. Explain your answer.

13 A shop has two jars of jelly beans. Jar A contains red jelly beans and green jelly beans in the ratio of 1 : 2. Jar B contains red and orange beans in the ratio 3 : 4.

There are three times as many jelly beans in jar A as there are in jar B.

All the beans are put into jar A and mixed thoroughly.

A bean is taken at random from jar A. Work out the probability that it is an orange jelly bean.

14 Four brothers, David, Malcolm, Brian and Kevin, regularly run races against each other in the park.

The chance of:

   David winning the race is 0.3

   Malcolm winning the race is $\frac{1}{5}$

   Brian winning the race is 45%.

What is the chance of Kevin winning the race?

15 A roulette wheel has 37 spaces for the ball to land on. The spaces are numbered 0–36. I always bet on a prime number.

If I play the game all evening and plan to play 100 times, how many times would I expect to win on the roulette table?

16 A head teacher is told that the probability of any student being left-handed is 0.14.

How can she work out how many of her students she should expect to be left-handed?

17 There are 53 students in a group, all studying English or mathematics or both. 30 of them study English and one-third of these also study mathematics. A student is chosen at random from these students. What is the probability that they study mathematics?

18 Andrew made a six-sided spinner.

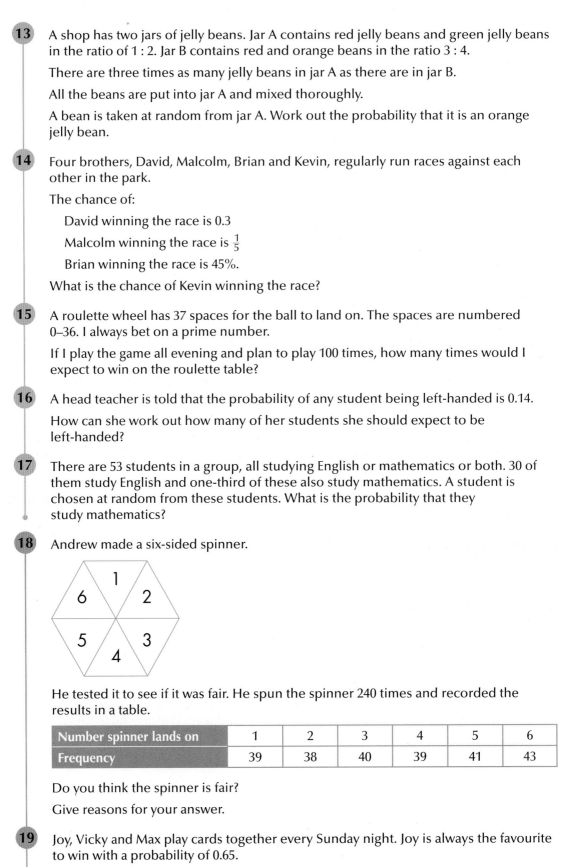

He tested it to see if it was fair. He spun the spinner 240 times and recorded the results in a table.

Number spinner lands on	1	2	3	4	5	6
Frequency	39	38	40	39	41	43

Do you think the spinner is fair?

Give reasons for your answer.

19 Joy, Vicky and Max play cards together every Sunday night. Joy is always the favourite to win with a probability of 0.65.

There were 52 Sundays in the year and Vicky won 10 times.

How many times in the year would you expect Max to have won?

20 Two of these six people are to be chosen for a job.

Anna, Ben, Chloe, Clara, Ciaran, Daniel

  a List all of the possible pairs (there are 15 altogether).

  b What is the probability that the pair of people chosen will:

   i both be female

   ii both be male

   iii both have the same initial

   iv have different initials?

  c Which of these pairs of events are mutually exclusive?

   i Picking two women and picking two men.

   ii Picking two people of the same sex and picking two people of opposite sex.

   iii Picking two people with the same initial and picking two men.

   iv Picking two people with the same initial and picking two women.

  d Which pair of mutually exclusive events in part **c** is also exhaustive?

21 **a** Explain the key features of mutually exclusive and independent events when they are shown on a probability tree diagram.

  b Explain why the probabilities on each set of branches have to sum to 1.

  c How can you tell from a completed probability tree diagram whether the question specified with or without replacement? You can use an example to help your explanation.

  d What strategies do you use to check the probabilities on your probability tree diagram are correct?

22 The weather forecast says the probability of rain on Thursday is 25% and on Saturday is 48%. What is the probability that it will rain on just one of the days?

23 Shehab walked into his local supermarket and saw a competition.

Roll two dice!
Score a total of two
and win a £20 note.
Only 50p a go.

  a What is the probability of winning a £20 note?

  b How many goes should he have in order to expect to win at least once?

  c If he had 100 goes, how many times could he expect to win?

24 A fair spinner has three equal sections. The spinner is spun twice.

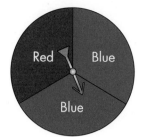

Show that the probability of scoring two blues is the same as scoring one red and one blue.

25 Thomas takes a driving test which is in two parts. The first part is a theory test. He has a 0.4 chance of passing this. The second part is practical. He has a 0.5 chance of passing this.

   a Draw a probability tree diagram covering passing or failing the two parts of the test.

   b What is the probability that he passes both parts?

26 In the game 'Rushdown', you are dealt two cards from a normal pack of cards. If you are dealt any two of the numbers 6 or 7 or 8, you have been dealt a 'Tango'.

What is the probability of being dealt a 'Tango'? Give your answer to three decimal places.

27 Anne regularly goes to London by train.

The probability of the train arriving in London late is 0.08.

The probability of the train being early is 0.02.

The probability of it raining in London is 0.3.

What is the probability of:

   a Anne getting to London on time and it not raining

   b Anne travelling to London three days in a row and it raining every day

   c Anne travelling to London five days in a row and not being late at all?

⑦Hints and tips

Question	Hint
3	Think about how many times it will happen in every 10.
6	Think about the likelihood of each of these outcomes happening.
7	Mutually exclusive outcomes cannot happen at the same time.
13	There are 7 parts in jar B so if there are three times as many parts in jar A, there will be 21 parts split in the ratio 1 : 2.
17	Use a Venn diagram and remember P(English only) + P(E∩M) + P(Mathematics only) = 1
26	Draw the sample space for this event.

Worked exemplars

 1 400 tickets are sold for a raffle. There is only one prize.

Mr Raza buys five tickets for himself and sells another 40.

Mrs Raza buys 10 tickets for herself and sells another 50.

Mrs Hewes just sells 52 tickets.

a What is the probability of:

   i Mr Raza winning the raffle

   ii Mr Raza selling the winning ticket to someone else?

b What is the probability of either Mr or Mrs Raza selling the winning ticket to someone else?

c What is the probability of Mrs Hewes not selling a winning ticket?

d Which of these three people has the greatest chance of either winning the raffle or selling the winning ticket? Give a reason for your answer.

Give all probabilities as fractions in their simplest form.

This is a mathematical reasoning question and so you must communicate your method clearly. Do not just write down probabilities without some explanation.	
a **i** $\dfrac{5}{400} = \dfrac{1}{80}$ **ii** $\dfrac{40}{400} = \dfrac{1}{10}$	Remember to cancel the fractions and make sure you read the information given in the question.
b $\dfrac{40}{400} + \dfrac{50}{400} = \dfrac{90}{400}$ $= \dfrac{9}{40}$	Remember that 'or' means add the two separate events. When adding the two fractions, they must have the same denominator.
c $1 - \dfrac{52}{400} = \dfrac{348}{400}$ $= \dfrac{87}{100}$	This is a question where you are working out the probability of an event not happening.
d P(Mr Raza either winning the raffle or selling the winning ticket) $= \dfrac{45}{400}$ P(Mrs Raza either winning the raffle or selling the winning ticket) $= \dfrac{60}{400}$ P(Mrs Hewes either winning the raffle or selling the winning ticket) $= \dfrac{52}{400}$ Mrs Raza has the greatest chance as $\dfrac{60}{400}$ is the largest fraction.	As you will need to compare fractions to solve the problem, there is no need to cancel the three probability fractions. Make sure you state your conclusion clearly and give a reason.

2 Susie is taking a driving test. The test is made up of two parts, a practical and a theory. She is told that the probability of passing only one of the two parts is 0.44 and the probability of passing the practical part of the test is 0.8.

    a If P(passing the theory part) = x, write down P(not passing the theory part).

    b Draw a probability tree diagram to show this information.

    c Set up an equation, in terms of x, to calculate the probability of passing the theory part.

This is a problem-solving question. You need to process the problem into a series of algebraic steps.	
a $1 - x$	Remember, P(event not happening) = 1 – P(event happening).
b	Remember to multiply the probabilities on the branches for each outcome.

c   P(only pass one part of the test)
   = P(PF) + P(FP)
   = $0.8(1 - x) + 0.2x$
So $0.8 - 0.8x + 0.2x = 0.44$
     $0.8 - 0.6x = 0.44$
        $0.6x = 0.8 - 0.44$
          $x = \dfrac{0.8 - 0.44}{0.6}$
          $x = 0.6$
So the probability of passing the theory part is 0.6.
	Use the probability tree diagram to work out P(only pass one part of the test).
	
	There are two outcomes on the diagram – PF and FP. Remember to add the two probabilities.
	
	This is the equation to solve.
	
	Rearrange the equation to calculate x.

6 Statistics

1 Pat measured the heights, to the nearest centimetre, of all the students in her class. Her data is given below.

a Make up three questions that can be answered using this information.

147	153	146	138	151	142	139	131	144	127	143	145	140	143
153	141	150	137	136	125	136	143	135	147	153	146	138	151
142	139	131	144	127	143	145	140	143	153	141	150	137	136
125	136	140	131	147	154	142							

b How would you represent the information? What makes the information easy or difficult to represent?

2 When drawing a pie chart, what information do you need to calculate the size of the angle for each category?

3 Azam timed how long each patient waited in a doctor's surgery one evening.

The following is his record in minutes.

28	24	31	11	31	6	24	5	16	17	10	29	7	15	20
26	22	19	27	13	24	27	15	32	8	4	38	19	14	12
33	22	34	21	25	9	25	5	18	29	14	13	23	30	8
35	20	29												

Draw a grouped frequency table for this data.

Explain why you chose the class intervals you used.

4 Hannah wants to survey the prices charged for soft drinks in her neighbourhood.

She says, 'I will make a frequency table with the class intervals 1p–40p, 40p–60p and 60p–£1.'

Give two reasons why these class intervals may be a problem.

5 The two pie charts show sales in two fruit and vegetable shops. One is in town and the other is in a small village.

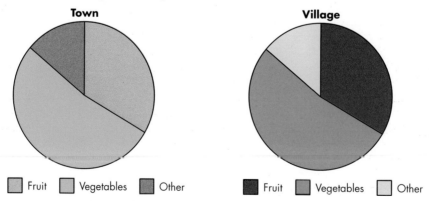

Which of these statements can you make using the pie charts? Explain why.

A: More people buy vegetables in the village than in the town.

B: The proportion of people buying fruit in the town is greater than in the village.

6 When plotting a graph to show the number of people attending cricket matches at a cricket ground, Kevin decides to start the vertical scale on his graph at 18 000. Explain why he might do this.

7 **a** List a small set of data that has a mode of 6. How did you do it?

  b List a small set of data that has a mode of 6 and a range of 15. How did you work this out?

  c Work out two different small sets of data that have the same mode and range. How did you do it?

  d Write a data set with a mean of 4 and a median of 3.

8 A tennis club organises a tournament for its junior members each month. The bar chart shows the number of boys and girls in the tournament over four months.

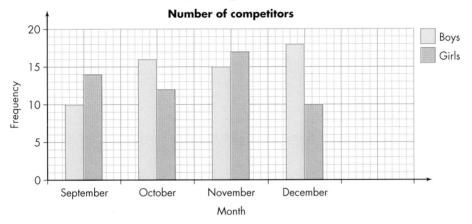

Number of competitors

  a In two of the months there were the same number of competitors in the tournament. Which two months?

  b There are 25 boys and 18 girls in the tennis club. In January, $\frac{2}{5}$ of the boys and $\frac{2}{3}$ of the girls took part in the tournament. How many boys and girls took part? Was this a good month for the club? Explain your answer.

9 Noah carried out a survey of how many men and women at his company wear cycle helmets when cycling to work. He used the information to draw these pie charts.

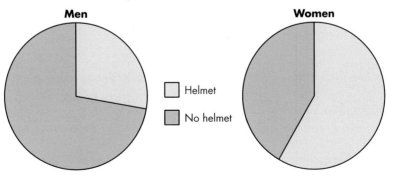

Men   Women

Helmet
No helmet

  a Mia looks at the pie charts and says: 'The pie charts show that more woman than men wear helmets.'

  Is she right? Explain your answer.

  b Noah chose to use pie charts for his data. Was it the best choice? Explain your answer.

10 This table shows the number of accidents involving buses that happened in the town of Redlow over a six-year period.

Year	2009	2010	2011	2012	2013	2014
No. of accidents	13	17	14	18	13	19

  a Draw a pictogram for this data.

  b Draw a bar chart for this data.

  c Which diagram would you use if you were going to write an article for the local newspaper trying to show that bus drivers need to take more care? Explain your answer.

11 Use the data-handling cycle to describe how you would test each of the following hypotheses. State in each case whether you would use primary or secondary data.

  a January is the coldest month of the year.

  b Girls are better than boys at estimating masses.

  c More men go to cricket matches than women.

  d A TV show is watched by more women than men.

  e The older you are the more likely you are to go ballroom dancing.

12 Godwin was asked to create a stem-and-leaf diagram from some numerical data, but he said, 'It is impossible to do this sensibly!'

Give an example of 10 items of numerical data that could not sensibly be put into a stem-and-leaf diagram.

13 A nationwide survey asked where people think the friendliest people in England are. The results are shown in the pie chart.

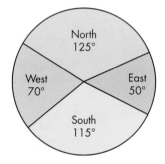

  a Write a statement or question for this graph using one or more of these words:

    total, range, mode, fraction, percentage, proportion, probability

  b Are there any words that you would find difficult to use for this example? Is so, explain why.

14 The graph shows the sales of women's shoes one day at a local shop.

  a What is wrong with showing this data on a line graph? What should it be plotted as?

  b Give an example of a situation where you would use a line graph.

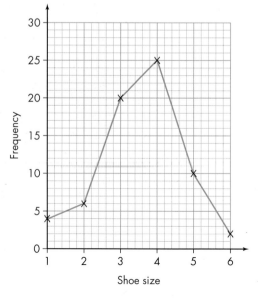

15 Alex is doing some research into life expectancy. His hypothesis is that everyone lives longer now than they did in 1960. He has produced these two graphs showing the same set of data on life expectancy.

Which chart is most helpful in testing the hypothesis? Why?

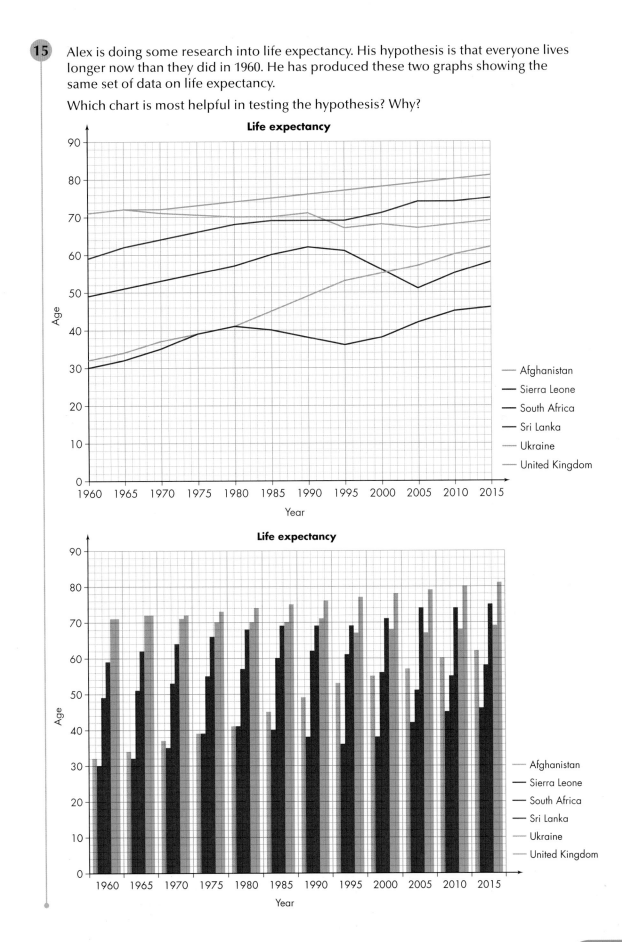

16 Doubling each number in a data set doubles the mean. Explain why this statement is true or false.

17 Asma travelled to Manchester on many days throughout the year.

The table shows how many days she travelled in each week.

Days	0	1	2	3	4	5
Frequency (no. of weeks)	17	2	4	13	15	1

Explain how you would work out the median number of days that Asma travelled in a week to Manchester.

18 The profit, to the nearest pound, made each week by a tea shop is shown in the table.

Profit	£0–£200	£201–£400	£401–£600	£601–£800
Frequency	15	26	8	3

Explain how you would estimate the mean profit made each week.

19 You are asked to draw a pie chart representing the different breakfasts that students have on one morning. What data would you need to get to do this and how would you get it?

20 **a** Write down three numbers that have a range of 3 and a mean of 3.

b Write down three numbers that have a range of 3, a median of 3 and a mean of 3.

21 A class of students took a test. The teacher said the average for the test was 32, but a student said the average was 28.

They were both correct. Explain how this could be the case.

22 Describe what you would expect a scatter graph to look like when it shows positive correlation.

23 The two-way table shows the part-time earnings of a group of students during one summer break.

Earnings per week	Male	Female
$£0 \leqslant E < £50$	4	1
$£50 \leqslant E < £100$	4	1
$£100 \leqslant E < £150$	11	4
$£150 \leqslant E < £200$	24	14
$£200 \leqslant E < £250$	16	10
$E \geqslant £250$	2	1

a What percentage of the male students earned between £100 and £150 per week?

b What percentage of the female students earned between £100 and £150 per week?

c Which sex has the greater estimated mean earnings? Explain how you would calculate this without doing the actual calculation.

24 Mr Phillips carried out a survey on the number of text messages received by students in his class on one day.

The graph shows the results for the boys.

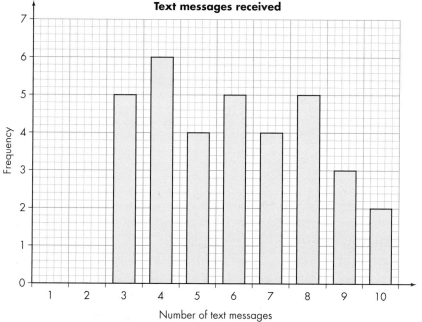

The mean number of text messages for the girls was 6.2.

Mr Phillips concluded that girls text more than boys. Would you agree with him? Explain your answer fully.

25 The table shows life expectancy (in years) for Ukraine and the United Kingdom from 1960 to 2010.

a Use the information to compare the data.

b Use the data to predict life expectancy in 2025. Which country is it easier to predict? Explain why.

Year	1960	1965	1970	1975	1980	1985	1990	1995	2000	2005	2010
Ukraine	71.06	71.63	70.87	70.28	69.7	69.88	70.46	66.79	67.79	67.31	68.18
United Kingdom	71.1	71.61	72.00	72.78	73.72	74.68	75.89	76.78	78.04	79.3	80.08

26 The table shows the marks for 10 students in their mathematics and music examinations.

Student	Mathematics	Music
Alex	52	50
Ben	42	52
Chris	65	60
Don	60	59
Ellie	77	61
Faisal	83	74
Gary	78	64
Hazel	87	68
Irene	29	26
Jez	53	45

a Plot the data on a scatter diagram. Plot the mathematics score on the x-axis and the music score on the y-axis.

b Draw the line of best fit.

c One of the students was ill when they took the mathematics examination. Which student was it most likely to be?

d Another student, Kris, was absent for the music examination but scored 45 in mathematics. What mark would you expect him to have got in music?

e Another student, Lex, was absent for the mathematics examination but scored 78 in music. What mark would you expect him to have got in mathematics?

27 The table shows the results of four different opinion polls recorded on one day just before a general election.

Combine all four polls into a single pie chart that represents the opinion of the UK population on this day.

Poll	Labour	Conservative	Liberal Democrat	Other
Poll 1	32%	37%	19%	12%
Poll 2	33%	37%	21%	9%
Poll 3	30%	35%	21%	14%
Poll 4	30%	37%	20%	13%

28 A number between 1 and 20 is written on each of four cards. The mean of the four numbers is twice the mode. Work out a possible set of numbers.

29 Here are the scores of class 8P in a recent mathematics test. One of the girls says this shows that girls are better at mathematics than boys because two girls got over 90% and the lowest score was from a boy.

Provide a counterargument to this to show that the boys are better at mathematics.

Girls' scores %	32	51	24	47	95	56	91	69	24	33	47	52	64	50			
Boys' scores %	46	68	87	56	54	13	87	68	82	46	57	64	38	82	82	66	75

30 Here is a stem and leaf diagram showing the results of the top 15 male and female runners from a 10 km race.

Male					Stem	Female			
				38	33				
			57	45	35				
		55	33	17	36				
58	38	36	20	01	37				
	42	25	21	13	38	40			
					39	08	08		
					40	18			
					41	15	42		
					42	18	46	54	59
					43	53	58		
					44				
					45	33			
					46	34	37		

Key Male: 40 | 33 represents 33 minutes and 40 seconds
Female: 33 | 40 represents 33 minutes and 40 seconds

a Compare the distributions for male and female runners.

b Write a few sentences to summarise your findings for a national running magazine.

31 An opinion poll used a sample of 200 voters in one area. 112 people said they would vote for Party A. There are 50 000 voters in the area.

a How many voters would you expect to vote for Party A if they all voted?

b The poll is accurate to within 10%. Can Party A be confident of winning?

32 Sandila's school has 1260 students and there are 28 students in her class. A survey is carried out using students sampled from the whole school. Four boys and three girls in Sandila's class are chosen to take part in the survey.

Estimate how many students in the whole school are in the sample.

33 You are asked to conduct a survey at a concert where the attendance is approximately 8000.

Explain how you could create a sample of the audience.

34 An aunt tells you that the local train service is not as good as it used to be.

a How could you check if this is true?

b Decide which data would be relevant to the enquiry and the possible sources. How might you collect this data?

35 A coffee stain removed four numbers (in two columns) from the following frequency table of eggs laid by 20 hens one day.

The mean number of eggs laid was 2.5.

What could the missing four numbers be?

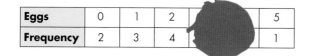

Eggs	0	1	2			5
Frequency	2	3	4			1

36 A hospital has to report the average waiting time for patients in the Accident and Emergency department. A survey was carried out to see how long patients waited before seeing a doctor.

The table summarises the results for one shift. The times are rounded to the nearest minute.

Time (minutes)	0–10	11–20	21–30	31–40	41–50	51–60	61–70
Frequency	1	12	24	15	13	9	5

a How many patients were seen by a doctor in this shift?

b Estimate the mean waiting time per patient.

c Which average would the hospital use to report the average waiting time?

d What percentage of patients did the doctors see within one hour?

37 Marion is writing an article on health for a magazine. She asked a sample of people the question: 'When planning your diet, do you consider your health?'

The pie chart shows the results of her survey.

a Can you tell how many people there were in the sample? Give a reason for your answer.

b Write a brief summary of the article Marion could write using this data.

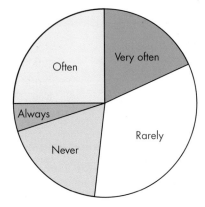

38 The frequency polygon shows the times that a number of people waited at the bus stop before their bus came one morning.

Dan says, 'Most people spent five minutes waiting.' Explain why this is incorrect.

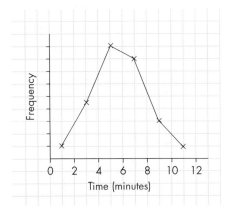

39 Andrew is asked to create a histogram. Explain to Andrew how he can work out the height of each bar on the frequency density scale.

40 Three dancers were hoping to be chosen to represent their school in a competition.

They had all been involved in previous competitions.

The table shows their scores in recent competitions.

Kathy	8, 5, 6, 5, 7, 4, 5
Connie	8, 2, 7, 9, 2
Evie	8, 1, 8, 2, 3

The teachers said they would be chosen by their best average score. Which average would each dancer prefer to be chosen by?

41 Jasmin has just graduated from university. Two different companies are trying to convince Jasmin to join them as an employee. Company A has just sent her these figures.

Mean salary from firm A is £32 000 to the nearest thousand.

Mean salary from firm B is £28 000 to the nearest thousand.

Therefore it is best to have a job with firm A.

However, Jasmin has been talking to a graduate advisor from her university. The advisor showed Jasmin this table of salaries for the two companies.

Imagine you are the graduate advisor. What advice would you give to Jasmin about the salaries of the two firms? Justify your recommendation to Jasmin.

Salaries from firm A	Salaries from firm B
£86 000	£45 000
£62 000	£36 000
£23 000	£26 000
£23 000	£26 000
£18 000	£26 000
£18 000	£26 000
£18 000	£22 000
£18 000	£22 000
£18 000	£22 000

42 Produce sets of grouped data with:

a an estimated range of 26

b an estimated median of 46

c an estimated median of 22.5 and an estimated range of 62

d an estimated mean of 36 (to one decimal place).

43 How would you make up a set of data with a median of 10 and an interquartile range of 7?

44 Connie planted some tomato plants and kept them in the kitchen, while her husband Harold planted some in the garden. After the summer, they compared their tomatoes.

	Connie	Harold
Mean diameter (cm)	1.9	4.3
Mean number of tomatoes per plant	23.2	12.3

Use the data in the table to explain who had the better crop of tomatoes.

45 **a** Which one of the following statements is more precise: 'My hypothesis is true.' or 'There is strong evidence to support my hypothesis.' Why?

b Why might it not be the case that:

i your hypothesis is true if you have found some evidence to support it

ii you have failed if your hypothesis appears to be flawed?

⑦Hints and tips

Question	Hint
7d	Think of a number that has a factor of 4. This number will be the sum of your data values. Then think about how many times 4 goes into this number and that will tell you how many pieces of data you need in your data set. Then all you have to do it put 3 in the middle and make sure the numbers either side of it will add up to the sum of your data values.
24	Think carefully about what he can actually tell about the students from the data you are given.
25	Use a suitable graph to help your comparison.
34	How can 'good' be defined? Frequency of service, cost of journey, time taken, factors relating to comfort, access? How does the frequency of the train service vary throughout the day and week?

Worked exemplars

1 The dual bar chart shows the number of people who attended an evening class lesson.

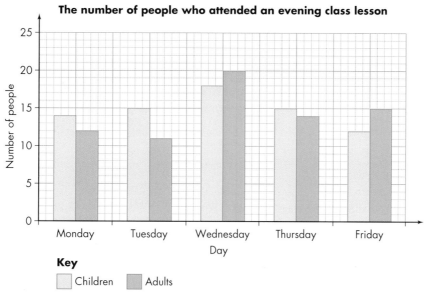

The number of people who attended an evening class lesson

Key
Children   Adults

a How many adults attended the lesson on Tuesday?

b On which day did most children attend the lesson?

c How many people attended the lesson on Friday?

d Louise said: 'More adults than children attended the five lessons.'

Is she correct? Give a reason for your answer, clearly showing your working.

This question requires you to interpret and draw conclusions from mathematical information.	
a 11	Make sure you read the correct value carefully.
b Wednesday	This is the day with the highest bar for the children. Make sure you give the day. The answer is not 20.
c 27	Add together the heights of the bars for Friday.
d No, because 74 children attended the five lessons but only 72 adults attended.	Lay out your working as shown in the table.

	Mon	Tue	Wed	Thu	Fri	Total
C	14	15	18	15	12	74
A	12	11	20	14	15	72

2 These are the masses, in kilograms, of a rowing boat crew.

91, 81, 89, 91, 79, 85, 87, 45

a Write down the modal mass.

b Work out the median mass.

c Calculate the mean mass.

d Which average best describes the data? Give a reason to support your answer.

In part **d** you need to evaluate your answers to parts **a** to **c** in relation to a set of data. First, work out the answers for parts **a** to **c**.	
a 91 kg	The mode is the most common mass.
b Masses in order: 45, 79, 81, 85, 87, 89, 91, 91 Median = 86 kg	Start by putting the masses in order. The middle pair is 85 and 87, so the median is the mass that is halfway between them.
c Mean = $\dfrac{91 + 81 + 89 + 91 + 79 + 85 + 87 + 45}{8}$ $= \dfrac{648}{8}$ $= 81$ kg	Add up all the masses and divide by the number in the crew. This gives $648 \div 8$.
d The median, as it avoids using the outlier mass of 45 kg.	This is an evaluation question so you need to interpret the three averages from parts **a** to **c** in the context of the given problem.

3 Roger and Brian keep a record of their scores in eight games of darts.

Game number	1	2	3	4	5	6	7	8
Roger	45	60	142	74	48	54	89	64
Brian	37	180	120	46	72	80	97	48

a What is the range of scores for each player?

b What is the mean score for each player?

c Which player is the more consistent and why?

d Who would you say is the better player and why?

This question requires you to show your skills in mathematical reasoning, which means that you should show how you reach your answer.	
a Roger's range = 142 – 45 = 97 Brian's range = 180 – 37 = 143	The range is the difference between the highest value and the smallest value.
b Roger's mean is 72 and Brian's mean is 85.	Add up all the scores for each player and divide by the number of games. Roger's mean = $576 \div 8 = 72$ Brian's mean = $680 \div 8 = 85$
c Roger, because his range is smaller.	You are drawing a conclusion from the information you have worked out.
d Brian, because he has a higher mean score.	You are drawing a conclusion from the information you have worked out.

4 The table shows the numbers of learners at each level for two practice driving tests, theory and practical.

| | Level | | | | | |
	Excellent	Very good	Good	Pass	Fail	Total number of learners
Theory	208	888	1032	696	56	2880
Practical	240	351	291	108	90	1080

a Represent the data for each of the two practice tests in a separate pie chart.

b On which test (theory or practical) do you think learners did better overall? Give a reason to justify your answer.

This question requires you to communicate your mathematical skills. Take care to interpret the information you are given accurately and show how you use it.

a Theory

Level	Frequency	Calculation	Angle
Excellent	208	$\frac{208}{2880} \times 360°$	26°
Very good	888	$\frac{888}{2880} \times 360°$	111°
Good	1032	$\frac{1032}{2880} \times 360°$	129°
Pass	696	$\frac{696}{2880} \times 360°$	87°
Fail	56	$\frac{56}{2880} \times 360°$	7°

Work out the angle for each level.

Remember to check that all the angles add up to 360°.

Take care to interpret the information you are given accurately.

Levels for driving test: theory

Remember to label the pie chart. You do not need to show the angles.

Practical

Level	Frequency	Calculation	Angle
Excellent	240	$\frac{240}{1080} \times 360°$	80°
Very good	351	$\frac{351}{1080} \times 360°$	117°
Good	291	$\frac{291}{1080} \times 360°$	97°
Pass	108	$\frac{108}{1080} \times 360°$	36°
Fail	90	$\frac{90}{1080} \times 360°$	30°

Work out the angle for each level.

Remember to check that all the angles add up to 360°.

Take care to interpret the information you are given accurately.

Levels for driving test: practical

Again, remember to label the pie chart. You do not need to show the angles.

b Overall the learners did better on the practical as 55% obtained Excellent or Very good, whereas only 38% obtained Excellent or Very good on the theory.

You must justify your answer. This is for interpreting and communicating the information accurately.

Glossary

3D shape   A shape with three dimensions, length, width and height.

acceleration   The rate at which the velocity of a moving object increases.

acute-angled triangle   A triangle in which all the angles are acute.

adjacent side   The side that is between a given angle and the right angle, in a right-angled triangle.

allied angles   Interior angles that lie on the same side of a line that cuts a pair of parallel lines; they add up to 180°.

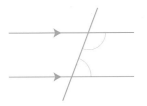

alternate angles   Angles that lie on either side of a line that cuts a pair of parallel lines; the line forms two pairs of alternate angles and the angles in each pair are equal.

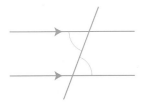

angle bisector   A line or line segment that divides an angle into two equal parts.

angle of depression   The angle between the horizontal line of sight of an observer and the direct line to an object that is viewed from above.

angle of elevation   The angle between the horizontal line of sight of an observer and the direct line to an object that is viewed from below.

angle of rotation   The angle through which an object is rotated, to form the image.

angles around a point   The angles formed at a point where two or more lines meet; their sum is 360°.

angles on a straight line   The angles formed at a point where one or more inclined (sloping) lines meet on one side of a straight line; their sum is 180°.

annual rate   A rate, such as interest, that is charged over a period of a year.

apex   The top point of a pyramid where all the edges of the sloping sides meet.

approximate   A value that is close but not exactly equal to another value, which can be used to give an idea of the size of the value; for example, a journey taking 58 minutes may be described as 'taking approximately an hour'; the ≈ sign means 'is approximately equal to'.

approximation   A calculated guess.

arc   Part of the circumference of a circle.

arithmetic sequence   A sequence of numbers in which the difference between one term and the next is constant.

average   A single value that represents a typical value for the whole set of data. The most common averages are the mode, the median and the mean.

average speed   The result of dividing the total distance travelled by the total time taken for a journey.

balancing   Doing the same thing to both sides of an equation.

bar chart   A diagram that is a series of bars or blocks of the same width to show frequencies.

bearing   The angle measured from north to define a direction.

best buy   The price that gives best value for money, the greatest quantity for the least price.

better value   The choice that gives more product per pound or penny.

bias   The property of a sample being unrepresentative of the population; for example, a dice may be weighted so that it gives a score of 5 more frequently than any other score.

bisect   Cut exactly in half.

cancel   When the numerator and denominator of a fraction have a common factor, this can be divided into both values to reduce the fraction to a fraction in its lowest terms.

capacity   The amount a container can hold.

categorical   Data that has non-numerical values.

centilitre (cl)   A measurement of capacity. A hundredth of a litre.

centre of enlargement   The point, inside, outside or on the perimeter of the object, on which an enlargement is centred; the point from which the enlargement of an object is measured.

centre of rotation   The point about which an object or shape is rotated.

chord   A straight line joining two points on the circumference of a circle.

circumference   The perimeter of a circle; every point on the circumference is the same distance from the centre, and this distance is the radius.

class interval   The range of a group of values in a set of grouped data.

coefficient   A number written in front of a variable in an algebraic term; for example, in $8x$, 8 is the coefficient of x.

column method   A method for multiplying large numbers, in which you multiply the units, tens and hundreds separately, then add the products together. This is also known as the *traditional method*.

common factor   A factor that divides exactly into two or more numbers; 2 is a common factor of 6, 8 and 10.

common unit   To enable you to compare quantities or simplify ratios, they must be expressed in the same or common units; for example, 2 m : 10 cm = 200 cm : 10 cm = 20 : 1.

complement   An event that does not happen. The probability that event A does not happen is written as P(A′).

composite bar chart   A bar chart where each bar compares sets of related data.

compound interest   Interest that is paid on the amount in the account; after the first year interest is paid on interest earned in the previous years.

compound shape   A shape made up of a combination of two or more shapes.

congruent   Exactly the same shape and size.

consecutive   Numbers that follow each other continuously eg 3, 4, 5, 6 are consecutive whole numbers, 4, 9, 16, 25 are consecutive square numbers.

consistency   A way of comparing two or more sets of data. The data set with the smallest range is said to be more consistent.

constant of proportionality   If two variables are in direct proportion, you can write an equation, $y = kx$; if they are in inverse proportion, you can write $xy = k$. In either case, k is the constant of proportionality.

constant term   A term that has a fixed value; in the equation $y = 3x + 6$, the values of x and y may change, but 6 is a constant term.

construct   Draw a shape by means of a ruler and a pair of compasses.

continuous data   Data, such as mass, length or height, that can take any value; continuous data has no precise fixed value.

conversion graph   A graph that can be used to convert from one unit to another.

correlation   A relationship or connection between two or more things showing how they vary in relation with each other, eg as the weather gets warmer, more ice creams are sold.

corresponding angles   Angles that lie on the same side of a pair of parallel lines cut by a line; the line forms four pairs of corresponding angles, and the angles in each pair are equal.

cosine   A trigonometric ratio related to an angle in a right-angled triangle, calculated as $\frac{adjacent}{hypotenuse}$.

cover-up   A method of solving equations by covering up one of the other terms.

cross-section   A cut across a 3D shape, or the shape of the face that is exposed when a 3D shape is cut. For a prism, a cut across the shape, perpendicular to its length.

cubic   An expression where the highest power of the variable is 3.

cylinder   A prism with a circular cross-section.

data collection sheet   Any way of collecting data for a survey in written form. See *tally chart*.

deceleration   The rate at which the velocity of a moving object decreases.

decimal fraction   Another name for a decimal. It implies that any decimal can be written as a fraction with a numerator of 10, 100, etc. For example $0.7 = \frac{7}{10}$ or $0.94 = \frac{94}{100} = \frac{47}{50}$.

decimal point   A symbol, usually a small dot, written between the whole-number part and the fractional part in a decimal number.

denominator   The bottom number in a fraction.

density   The mass of a substance divided by its volume.

diameter   A chord in a circle that passes through the centre.

difference   The result of a subtraction.

difference of two squares   An expression of the form $x^2 - y^2$: the terms are squares and there is a minus sign between them.

digit   Digits are the numbers from 0 to 9. 56 is a two-digit number.

direct proportion   A relationship in which one variable increases or decreases at the same rate as another; in the formula $y = 12x$, x and y are in direct proportion.

direct variation   Another name for direct proportion.

direction   The line along which a vector such as force, weight or velocity acts.

discrete data   Data that can only take certain values, such as a number of children; discrete data can only take fixed values.

distance   The length between two points. For example, the distance from A to B is 10 centimetres.

distance–time graph   A graph that represents a journey, based on the distance travelled and the time taken.

dual bar chart   A bar chart to compare two sets of related data.

edge   The line where two faces or surfaces of a 3D shape meet.

element   Any member of a set.

elevation   The view of a 3D shape when you view it from the side or the front.

eliminate   Given a pair of simultaneous equations with two variables, you can manipulate one or both equations to remove or eliminate one of the variables by a process of substitution, addition or subtraction.

enlargement   A transformation in which the object is enlarged to form an image.

equally likely   The outcomes of an event are equally likely when each has the same probability of occurring. For example, when tossing a fair coin the outcomes 'Heads' and 'Tails' are equally likely.

equation   A relation in which two expressions are separated by an equals sign with one or more variables. An equation can be solved to find one or more answers, but it may not be true for all values of x.

equidistant   At equal distances.

equilateral triangle   A triangle in which all the sides are equal and all the angles are 60°.

equivalent   Exactly the same as, usually used in 'equivalent fraction'.

equivalent fraction   Any fraction that can be made equal to another fraction by cancelling. For example $\frac{9}{12} = \frac{3}{4}$.

error interval   The interval within which a rounded value can lie. For example, if $x = 25$ to the nearest whole number, the error interval for x is $24.5 \leqslant x < 25.5$.

estimate   A calculated guess.

estimated mean   A mean that is estimated from grouped data, by multiplying the frequency by the mid-class value for each class, adding up the products and dividing by the total frequency.

event   Something that happens in a probability problem, such as tossing a coin or predicting the weather.

exhaustive   All possible outcomes of an event; the sum of the probabilities of exhaustive outcomes equals 1.

expand   Multiply out (terms with brackets).

expectation   Predicting the number of times you would expect an outcome to occur.

experiment   A method for collecting data, by carrying out a series of trials.

experimental data   Data that is collected by an experiment.

experimental probability   An estimate for the theoretical probability.

expression   A collection of numbers, letters, symbols and operators representing a number or amount; for example, $x^2 - 3x + 4$.

exterior angle   The angle formed outside a 2D shape, when a side is extended beyond the vertex.

extrapolation   To predict an outcome based on known facts or observations.

face   The area on a 3D shape enclosed by edges.

factor   A number that divides exactly (no remainder) into another number, for example, the factors of 12 are {1, 2, 3, 4, 6, 12}.

factor pair   A factor pair of a number is any pair of numbers whose product is the original number, for example, the factor pairs of 12 are 1×12, 2×6 and 3×4. Note that square numbers always have a number (the square root) that is its own 'pair', so 16 has a factor 'pair' of 4×4.

factor tree   A method of breaking down a number into its prime factors.

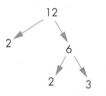

factorial   The product of the whole number n and all the whole numbers less than n down to 1. It is written as $n!$. For example $5! = 5 \times 4 \times 3 \times 2 \times 1 = 120$.

factorisation   The arrangement of a given number or expression into a product of its factors. (verb: factorise)

flow diagram   A diagram that splits a function into single steps.

foot (ft)   An imperial measurement of length. 3 feet = 1 yard

formula   A mathematical rule, using numbers and letters, which shows a relationship between variables; for example, the conversion formula from temperatures in Fahrenheit to temperatures in Celsius is: $C = \frac{5}{9}(F - 32)$.

frequency   The number of times each value occurs.

frequency table   A table that shows all the frequencies after all the data has been collected.

frequency tree diagram   A diagram that shows the frequencies when the probabilities of different events are known.

function key   See *shift key*.

function   An algebraic expression in which there is only one variable, often x.

gallon (gal)   An imperial measurement of volume. 1 gallon $\approx$ 4.55 litres

geometric sequence   A sequence in which each term is multiplied or divided by the same number, to produce the next term; for example, 2, 4, 8, 16, ... is a geometric sequence.

gradient   The slope of a line; the vertical difference between the coordinates divided by the horizontal difference.

gradient-intercept   A form for the equation of a line, written in terms of its gradient and the intercept on the vertical axis, $y = mx + c$ where m is the gradient and c is the y-intercept.

grid method   A method for multiplying numbers larger than 10, in which each number is split into its parts: for example, to calculate 158×67: 158 is 100, 50 and 8 and 67 is 60 and 7. These numbers are arranged in a grid and each part is multiplied by the others. This is also known as the *box method*.

×	100	50	8
60	6000	3000	480
7	700	350	56

```
   6 0 0 0
   3 0 0 0
     4 8 0
     7 0 0
     3 5 0
 +    5 6
 ─────────
 1 0 5 8 6
```

grouped data   Data arranged into smaller, non-overlapping sets, groups or classes, that can be treated as separate ranges or values, for example, 1–10, 11–20, 21–30, 31–40, 41–50; in this example there are equal class intervals.

grouped frequency table   A frequency table where the data has been collected using class intervals.

highest common factor   The largest number that is a factor common to two or more other numbers.

hypothesis   A statement that has to be proved true or false.

identity   Expressions either side of a $\equiv$ sign with one or more variables, which is true for all values; for example, $3(x + 2) \equiv 3x + 6$ is an identity.

image   The result of a reflection or other transformation of an object.

imperial   Units commonly used in Britain that are gradually being phased out as we change to the metric system.

improper fraction A fraction that has a numerator greater than the denominator.

inch (in) An imperial measurement of length. 12 inches = 1 foot.

included angle The angle made by two lines with a common vertex.

inclusive inequality An inequality such as ⩽ or ⩾.

index notation Expressing a number in terms of one or more of its factors, each expressed as a power.

inequality A statement that one expression is greater or less than another, written with the symbol > (greater than) or < (less than) instead of = (equals).

input The number that is put into a function.

intercept The point where a line cuts or crosses the axis.

interior angle The inside angle between two adjacent sides of a 2D shape, at a vertex.

interpolation To insert, estimate or find an intermediate value.

intersect To cross over.

intersection The 'overlap', the set of elements that occur in two or more sets.

inverse Going the other way.

inverse flow diagrams A flow diagram that shows the reverse process.

inverse operations An operation that reverses the effect of another operation; for example, addition is the inverse of subtraction, division is the inverse of multiplication.

inverse proportion A relationship between two variables in which as one value increases, the other decreases; in the formula $xy = 12$, x and y are in inverse proportion.

inverse variation Another name for inverse proportion.

invert Turn upside down. Usually used when dividing by a fraction. The calculation is turned into a multiplication by inverting the dividing fraction.

isometric grid A sheet with dots on to help draw a 3D representation.

isosceles triangle A triangle in which two sides are equal and the angles opposite the equal sides are also equal.

key A symbol that shows what each item represents.

like terms Terms in which the variables are identical, but the coefficients may be different; for example, $2ax$ and $5ax$ are like terms but $5xy$ and $7y$ are not. Like terms can be combined by adding their numerical coefficients so $2ax + 5ax = 7ax$.

line of best fit A straight line drawn on a scatter diagram where there is correlation, so that there are equal numbers of points above and below it; the line shows the trend of the data.

line segment A line joining two points.

linear graph A straight-line graph that represents a linear function.

linear sequence A sequence or pattern of numbers in which the difference between consecutive terms is always the same.

loci The plural of locus.

locus The path of a point that moves, obeying given conditions.

lowest common denominator When adding or subtracting fractions with different denominators, they must first be written with the same denominator. To avoid having to cancel this should be the lowest common multiple of the denominators.

lowest common multiple (LCM) The lowest number that is a multiple of two or more numbers; 12 is the lowest common multiple of 2, 3, 4 and 6.

magnitude The size of a quantity.

mass The amount of matter in an object.

metric The international standard for all measurement. Based on metres, kilograms and litres.

mid-class value The mid-point value of each class interval.

mirror line Another name for a line of symmetry.

mixed number A number made up of a whole number and a fraction, for example $3\frac{3}{4}$.

modal Any value that represents the mode.

modal group In grouped data, the class with the highest frequency.

multiple Any member of the times table, for example multiples of 7 are 7, 14, 21, 28, etc.

multiplication table   Any table with sets of numbers across the top and left hand side (usually 1 to 12) where the cells of the table are the products of the values in the top row and left hand column.

multiplier   A number that is used to find the result of increasing or decreasing an amount by a percentage.

multiply out   To multiply everything in a pair of brackets by the term in front of the brackets (or everything in one pair of brackets by everything in another pair of brackets).

mutually exclusive   Outcomes that cannot occur at the same time.

negative   Used in 'negative number' to mean a number less than zero.

negative coordinates   Coordinates such as (–2, 6), which contain one or more negative numbers.

negative correlation   A relationship between two sets of data, in which the values of one variable increase as the values of the other variable decrease.

net   A flat shape you can fold into a 3D shape.

no correlation   No relationship between two sets of data.

nth term   An expression in terms of n where n is the position of the term; it allows you to find any term in a sequence, without having to use a term-to-term rule.

numerator   The top number in a fraction.

object   The original or starting shape, line or point before it is transformed to give an image.

observation   A method for collecting data, by recording each item in a survey.

obtuse-angled triangle   A triangle containing an obtuse angle.

opposite side   The side that is opposite a given angle, in a right-angled triangle.

order of rotational symmetry   The number of times a 2D shape looks the same as it did originally when it is rotated through 360° about a central point. If a shape has no rotational symmetry, its order of rotational symmetry is 1, because every shape looks the same at the end of a 360° rotation as it did originally.

ounce (oz)   An imperial measurement of mass. 16 ounces = 1 pound

outcome   A possible result of an event in a probability experiment, such as the different scores when throwing a dice.

outlier   In a data set, a value that is widely separated from the main cluster of values.

output   The result when an input number is acted on by a function.

parabola   The shape of a quadratic curve.

parallel   Two lines which have the same gradient are called parallel lines.

parallelogram   A four sided shape where both pairs of opposite sides are parallel.

partition method   A method for long multiplication that requires the numbers to be written as separate digits across and down a grid. These are then multiplied and the values added diagonally to get the final answer. This is also known as 'Napier's Bones'.

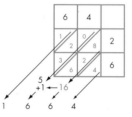

pattern   Numbers or objects that are arranged to follow a rule.

per cent   From Latin 'per centum' meaning out of a hundred.

percentage   Any fraction or decimal expressed as an equivalent fraction with a denominator of 100 but written with a percentage sign (%), for example $0.4 = \frac{40}{100} = 40\%$, $\frac{4}{25} = \frac{16}{100} = 16\%$.

percentage change   A change to a quantity, calculated as a percentage of the original quantity.

percentage decrease   A reduction or decrease to a quantity, calculated as a percentage of the original quantity.

percentage increase   An increase to a quantity, calculated as a percentage of the original quantity.

percentage loss   The loss on a financial transaction, calculated as the difference between the buying price and the selling price, calculated as a percentage of the original price.

percentage profit   The profit on a financial transaction, calculated as the difference between the selling price and the buying price, calculated as a percentage of the original price.

perpendicular bisector   A line that divides a given line exactly in half, passing through its midpoint at right angles to it.

perpendicular height   The shortest height from the base to a vertex.

pi (π)   The result of dividing the circumference of a circle by its diameter, represented by the Greek letter pi (π).

pictogram   A frequency table where the frequency for each type of data is shown by a symbol.

pie chart   A method of comparing discrete data. A circle is divided into sectors whose angles each represent a proportion of the whole sample.

place value   The position (place) of a digit in a number defines its value, so in 432.17, 4 represents 400 and 7 represents 7 hundredths.

plan   The view from directly above a solid shape.

polygon   A closed 2D shape with straight sides.

population   The complete data set in a survey.

position-to-term   A rule for generating a term in a sequence, depending on the position of the term within the sequence.

positive   Used in 'positive number' to mean a number greater than zero.

positive correlation   A relationship between two sets of data, in which the values of one variable increase as the values of the other variable increase.

pound (lb)   An imperial measurement of mass, approximately the same as a jar of jam.

power   The number of times you use a number or expression in a calculation; it is written as a small, raised number; for example, 2^2 is 2 multiplied by itself, $2^2 = 2 \times 2$ and $4^3 = 4 \times 4 \times 4$.

powers of 10   A number that is produced by multiplying 10 by itself repeatedly.

powers of 2   A number that is produced by multiplying 2 by itself repeatedly.

pressure   The amount of force exerted divided by the area on which that force acts.

primary data   Data you have collected yourself.

prime factor   Any factor of a number that is a prime number. For example the factors of 12 are {1, 2, 3, 4, 6, 12}. The prime factors of 12 are {2, 3}.

prime factorisation   Breaking a number down into a product consisting of prime factors only. See *product of prime factors*.

prime number   A number with only two factors, 1 and itself.

principal   The amount invested or lent.

prism   A 3D shape that has the same cross-section wherever it is cut perpendicular to its length.

probability   The chance of an event happening.

probability fraction   The probability of an event happening, written as a fraction. Probabilities can also be written as decimals or percentages.

probability scale   A scale between 0 and 1 that shows the probability of an event happening.

probability space diagram   A diagram that shows all the outcomes for two events.

product of prime factors   A number written as a product of its prime factors, for example $12 = 2 \times 2 \times 3$.

proper fraction   A fraction that is less than one, with the numerator less than the denominator.

pyramid   A 3D shape with a base and sides rising to form a single point.

Pythagoras' theorem   The rule that, in any right-angled triangle, the square of the hypotenuse is equal to the sum of the squares of the other two sides.

quadratic   Having terms involving one or two variables, and constants, such as $x^2 - 3$ or $y^2 + 2y + 4$ where the highest power of the variable is two.

quadratic expansion   Multiplying out two pairs of brackets, leading to a quadratic expression.

quadratic expression   An expression in which the highest power of any variable is 2, such as $2x^2 + 4$.

quadratic sequence   A sequence in which the first differences are not constant, formed from a quadratic rule.

quantity   A measurable amount of something that can be written as a number, or a number with appropriate units; for example, the capacity of a milk carton.

radius   A straight line joining the centre of a circle to any point on the circumference.

random   Chosen by chance, without looking; every item has an equal chance of being chosen.

Glossary

random sample   A sample in which every member of the population has an equal chance of being chosen.

range   The difference between the highest and lowest values for a set of data.

ratio   The ratio of A to B is a number found by dividing A by B. It is written as A : B. For example, the ratio of 1 m to 1 cm is written as 1 m : 1 cm = 100 : 1. Notice that the two quantities must both be in the same units if they are to be compared in this way.

rational number   A number that can be written as a fraction, for example, $\frac{1}{4}$ or $\frac{10}{3}$.

rearrange   Put into a different order, to simplify.

reciprocal   The result of dividing a number into 1, so 1 divided by the number is its reciprocal.

recurring decimal   A decimal number in which a digit or pattern of digits repeats for ever.

reflection   The image formed when a 2D shape is reflected in a mirror line or line of symmetry; the process of reflecting an object.

relative frequency   An estimate for the theoretical probability.

representative   A value that is typical for the whole set of data.

resultant vector   The result of combining two or more vectors.

right-angled triangle   A triangle in which one angle is 90°.

roots   The points on a graph where it crosses the x-axis.

rotation   A turn about a central point, called the centre of rotation.

rotational symmetry   A type of symmetry in which a 2D shape may be turned through 360° so that it looks the same as it did originally in two or more positions.

round   The process of giving an estimate by changing the number of significant figures.

sample   A selection taken from a larger data set, which can be researched to provide information about the whole population.

sample size   The number of items of data collected when doing a survey or testing a hypothesis.

sample space diagram   A diagram that shows all the outcomes of an experiment.

scalar   A quantity such as mass that has quantity but does not act in a specific direction.

scale   The number of squares that are used for each unit on an axis.

scale drawing   A drawing that represents something much larger or much smaller, in which the lengths on the image are in direct proportion to the lengths on the object.

scale factor   The ratio of the distance on the image to the distance it represents on the object; the number that tells you how much a shape is to be enlarged.

scalene triangle   A triangle in which all sides are different lengths.

scatter diagram   A graphical representation showing whether there is a relationship between two sets of data.

secondary data   Data that has been collected by someone else.

sector   A region of a circle, like a slice of a pie, enclosed by an arc and two radii.

segment   A chord will divide the circle into two segments, one each side of the chord.

sequence   A pattern of numbers that are related by a rule.

set   A collection of objects or elements.

shift key   The key on a calculator that enables you to use the alternative functions associated with the main keys.

significant figure   In the number 12 068, 1 is the first and most significant figure and 8 is the fifth and least significant figure. In 0.246 the first and most significant figure is 2. Zeros at the beginning or end of a number are not significant figures.

similar   Two shapes are similar if one is an enlargement of the other; angles in the same position in both shapes are equal to each other.

simple interest   Money that a borrower pays a lender, for allowing them to borrow money.

simplest form   A fraction written so that the numerator and denominator have no common factors.

simplify   To make an equation or expression easier to work with or understand by combining like terms or cancelling; for example:
$4a - 2a + 5b + 2b = 2a + 7b$, $\frac{12}{18} = \frac{2}{3}$, $5 : 10 = 1 : 2$.

simultaneous equations   Two equations that are both true for the same set of values for their variables.

sine   A trigonometric ratio related to an angle in a right-angled triangle, calculated as $\frac{\text{opposite}}{\text{hypotenuse}}$.

slant height   The length of the sloping side of a cone.

solution   The answer for an equation; the method of finding the answer.

speed   The rate at which an object moves. For example, the speed of the car was 40 miles per hour.

sphere   A 3D shape that is the locus of a point that moves a fixed distance from a given point, the centre; a 3D shape that has a circular cross-section whenever it is cut through its centre.

spread   A way of describing how a set of data is scattered for all the values. See *range*.

square number   A number formed when any integer is multiplied by itself. For example, $3 \times 3 = 9$ so 9 is a square number.

square root   A number that produces a specified quantity when multiplied by itself. For example, the square root of 16 is 4. Not all square roots are whole numbers. It uses the symbol $\sqrt{\ }$, so $\sqrt{25} = 5$, and $\sqrt{7} = 2.645\,751...$

standard form   A way of writing a number as $a \times 10^n$, where $1 \leqslant a < 10$ and n is a positive or negative integer.

standard index form   See *standard form*.

stone (st)   An imperial measurement of mass, approximately equal to 6 kilograms.

strict inequality   An inequality such as < or >.

subject   The variable on the left-hand side of the equals (=) sign in a formula or equation.

substitute   Replace a variable in an expression with a number and work out the value; for example, if you substitute 4 for t in $3t + 5$ the answer is 17 because $3 \times 4 + 5 = 17$.

subtend   The joining of the lines from two points giving an angle.

surface area   The total area of all of the surfaces of a 3D shape.

survey   A method of collecting data by asking questions or observing.

symbol   Symbols such as + and = are used to simplify expressions and equations.

systematic counting   If you wanted to work out how many times the digit 6 was written when writing down all the numbers from 200 to 300 you would use a systematic counting strategy; for example, 206, 216, … 296 is 10 times plus 260, 261, … 269 which is 10 times so the digit 6 will be written 20 times. Note that if the question was how many numbers between 200 and 300 contain the digit 6, the answer would be 19 as 266 would be counted only once.

tally chart   A data collection sheet where the data is collected using a tally.

tangent   1   A straight line that touches a circle just once.

　　　　　 2   A trigonometric ratio related to an angle in a right-angled triangle, calculated as $\frac{\text{opposite}}{\text{adjacent}}$.

term   1   A part of an expression, equation or formula. Terms are separated by + and – signs.

　　　　 2   A number in a sequence or pattern.

terminating decimal   A terminating decimal can be written down exactly. $\frac{33}{100}$ can be written as 0.33 so is a terminating decimal, but $\frac{1}{3}$ is 0.3333… with the 3s recurring forever.

term-to-term   The rule that shows what to do to one term in a sequence, to work out the next term.

theoretical probability   The exact or true probability of an event happening.

three-figure bearing   The angle from north clockwise, generally given as a three-digit figure.

time   A point or period of the day as measured in hours and minutes past midnight or noon. For example, the bus leaves at 10 am or the journey took 3 hours.

time series graph   A line graph where the horizontal axis represents time.

ton (T)   An imperial measurement of mass, approximately equal to a saloon car.

tonne (t)   A metric measurement of mass, approximately equal to a saloon car.

transformation   A change to a geometric 2D shape, such as a translation, rotation, reflection or enlargement.

translation   A movement along, up, down or diagonally on a coordinate grid.

trapezium   A four sided shape where only one pair of opposite sides are parallel.

tree diagram   A diagram that is used to calculate the probability of combined events happening. All the probabilities of each single event are written on the branches of the diagram.

trend   How data increases or decreases in a regular pattern.

trial   A single experiment in a probability experiment.

trigonometric functions   These are sine, cosine and tangent.

trigonometric ratios   These are sine $= \frac{\text{opposite}}{\text{hypotenuse}}$, cosine $= \frac{\text{adjacent}}{\text{hypotenuse}}$ and tangent $= \frac{\text{opposite}}{\text{adjacent}}$.

trigonometry   The study of the relationship between angles and sides in triangles.

turning point   Any point on a graph where the gradient is zero; for a quadratic graph this is the lowest or highest point.

two-way table   A table that records how two variables are linked.

unbiased   The property of a sample being representative of the population, so that any member of the population may be chosen.

union   The set of all the elements that occur in one or more sets.

unique factorisation theorem   This states that every integer greater than 1 is a prime number or can be written as a product of prime numbers.

unit cost   The cost of one unit, such as a kilogram, litre or metre, of something.

unitary method   A method of finding best value by finding the price per unit, or the quantity per pound or penny.

universal set   The set that contains all possible elements, usually represented by the symbol ξ.

value for money   When comparing costs or offers of the same item, which offer gives the least unit cost.

variable   A letter that stands for a quantity that can take various values.

vector   A quantity such as velocity that has magnitude and acts in a specific direction.

Venn diagram   A diagram that shows the relationships between different sets.

vertex   The point at which two lines meet, in a 2D or 3D shape.

vertical height   The height of the top vertex of a 3D shape, measured perpendicular to the base.

vertical line chart   Similar to a bar chart, but has vertical lines instead of bars.

vertically opposite angles   The angles on the opposite side of the point of intersection when two straight lines cross, forming four angles. The opposite angles are equal.

volume   The space taken up by a solid shape.

x-values   The first number in a pair of coordinates, the input of a function.

$y = mx + c$ The general equation of a straight line in which m is the gradient of the line and c is the intercept on the y-axis.

yard (yd)   An imperial measurement; 1 yard, the approximate distance from your fingertip to your nose when you stretch out your arm.

y-intercept   The point where a graph intersects the y-axis.

y-values   The second number in a pair of coordinates, the output of a function.

zero gradient   A line that is parallel to the horizontal axis has zero gradient.